Clara Erskine Clement Waters

An Outline History of Architecture for Beginners and Students

With Complete Indexes and Numerous Illustrations

Clara Erskine Clement Waters

An Outline History of Architecture for Beginners and Students
With Complete Indexes and Numerous Illustrations

ISBN/EAN: 9783337402013

Printed in Europe, USA, Canada, Australia, Japan

Cover: Foto ©berggeist007 / pixelio.de

More available books at **www.hansebooks.com**

AN OUTLINE HISTORY OF

ARCHITECTURE·

FOR

BEGINNERS AND STUDENTS

WITH

*COMPLETE INDEXES AND NUMEROUS
ILLUSTRATIONS*

BY

CLARA ERSKINE CLEMENT

AUTHOR OF "PAINTING FOR BEGINNERS AND STUDENTS," "SCULPTURE FOR BEGINNERS
AND STUDENTS," "HANDBOOK OF LEGENDARY AND MYTHOLOGICAL
ART," ETC., ETC.

FOURTH EDITION

NEW YORK
FREDERICK A. STOKES COMPANY
MDCCCXCIII

CONTENTS.

CHAPTER I.

CHAPTER II.

CHAPTER III.

CONTENTS.

LIST OF ILLUSTRATIONS.

ARCHITECTURE.

CHAPTER I.

ANCIENT OR HEATHEN ARCHITECTURE.

3000 B.C. TO A.D. 328.

A RCHITECTURE seems to me to be the most wonderful of all the arts. We may not love it as much as others, when we are young perhaps we cannot do so, because it is so great and so grand ; but at any time of life one can see that in Architecture some of the most marvellous achievements of men are displayed. The principal reason for saying this is that Architecture is not an imitative art, like Painting and Sculpture. The first picture that was ever painted was a portrait or an imitation of something that the painter had seen. So in Sculpture, the first statue or bas-relief was an attempt to reproduce some being or object that the sculptor had seen, or to make a work which combined portions of several things that he had observed ; but in Architecture this was not true. No temples or tombs or palaces existed until they had first taken form in the mind and imagination of the builders. and were created out of space and nothingness, so to speak. Thus Painting and Sculpture are imitative arts, but Architecture is a constructive art ; and while one may love pictures or statues more than the work of the architect, it seems to me that one must wonder most at the last.

We do not know how long the earth has existed, and in studying the most ancient times of which we have any accurate knowledge, we come upon facts which prove that men must have lived and died long before the dates of which we can speak exactly. The earliest nations of whose Architecture we can give an account are called heathen nations, and their art is called Ancient or Heathen Art, and this comes down to the time when the Roman Emperor Constantine was converted to Christianity, and changed the Roman Capitol from Rome to Constantinople in the year of our Lord 328.

The buildings and the ruins which still remain from these ancient times are in Egypt, Assyria, Persia, Judea, Asia Minor, Greece, Etruria, and Rome. Many of these have been excavated or uncovered, as, during the ages that have passed since their erection, they had been buried away from sight by the accumulation of earth about them. These excavations are always going on in various countries, and men are ever striving to learn more about the wonders of ancient days ; and we may hope that in the future as marvellous things may be revealed to us as have been shown in the past.

EGYPT.

As we consider the Architecture of Egypt, the Great Pyramid first attracts attention on account of its antiquity and its importance. This was built by Cheops, who is also called Suphis, about 3000 years before Christ. At that distant day the Egyptians seem to have been a nation of pyramid-builders, for even now, after all the years that have rolled between them and us, we know of more than sixty of these mysterious monuments which have been opened and explored.

Of all these the three pyramids at Ghizeh (Fig. 1) are best known, and that of Cheops is the most remarkable among them. Those of you who have studied the history of the wars of Napoleon I. will remember that it was near this spot that he fought the so-called Battle of the Pyramids, and that in addressing his soldiers he reminded them that here the ages looked down upon them, thus referring to the many years during which this great pyramid had stood on the border of the desert, as if watching the flight of Time and calmly waiting to see what would happen on the final day of all earthly things.

There have been much speculation and many opinions as to the use for which these pyramids were made, but the most general belief is that they were intended for the tombs of the powerful kings who reigned in Egypt and caused them to be built.

The pyramid of Cheops was four hundred and eighty feet and nine inches high, and its base was seven hundred and sixty-four feet square. It is so difficult to understand the size of anything from mere figures, that I shall try to make it plainer by saying that it covers more than thirteen acres of land, which is more than twice as much as is covered by any building in the world. Its height is as great as that of any cathedral spire in Europe, and more than twice that of the monument on Bunker Hill, which is but two hundred and twenty feet, and yet looks very high.

When it was built it was covered with a casing of stone, the different pieces being fitted together and polished to a surface like glass ; but this covering has been torn away and the stones used for other purposes, which has left the pyramid in a series of two hundred and three rough and jagged steps, some of them being two feet and a half in height, growing less toward the top, but not diminishing with any regularity. The top is now a platform thirty-two feet and eight inches square. Each traveller who ascends this pyramid has from one to four Fellahs or Arabs, who pull him forward or upward by his arms, or push him and lift him from behind, and finally drag him to the top (Fig. 2). When he thinks of all the weary months and days of the twenty years during which it is said that those who built it worked, cutting out the stone in the quarries, mov-

ing it to the spot where it was required, and then raising it to the great heights and fitting it all in place, he regards his fatigue in its ascent as a little thing, though at the time it is no joke to him.

Many of the pyramids were encased in stone taken from the Mokattam Mountains, which were somewhat

FIG. 2.—THE ASCENT OF A PYRAMID.

more than half a mile distant; but the pyramid of Cheops was covered with the red Syenite granite, which must have been quarried in the "red mountain," nearly five hundred miles away, near to Syene, or the modern Assouan. The interior of the pyramid is divided into chambers and passages (Fig. 3), which are lined with beautiful slabs of granite and con-

structed in such a way as to prove that at the remote time
in which the pyramids were built Egyptian architects and
workmen were already skilled in planning and executing
great works. Of the seventy pyramids known to have existed in those early days, sixty-nine had the entrance on the
north side, leaving but
a single exception to
this rule ; all of them
were situated on the
western side of the
River Nile, just on the
edge of the desert, be-
yond the strip of cul-
tivable ground which
borders the river.

Near the pyramids
there are numerous
tombs, which are built
somewhat like low
houses, having several
apartments with but
one entrance from the
outside. The walls of
these apartments are
adorned with pictures
similar to this one of a
poulterer's shop (Fig.
4) ; they represent the
manners and customs of
the ancient Egyptians
with great exactness.

FIG. 3.—VIEW OF GALLERY IN THE GREAT
PYRAMID.

The tombs at Beni-Hassan are among the most ancient
ruins of Egypt, and are very interesting (Fig. 5). They
were made between 2466 and 2266 B.C. They are on the
eastern bank of the Nile, and are hewn out of the solid

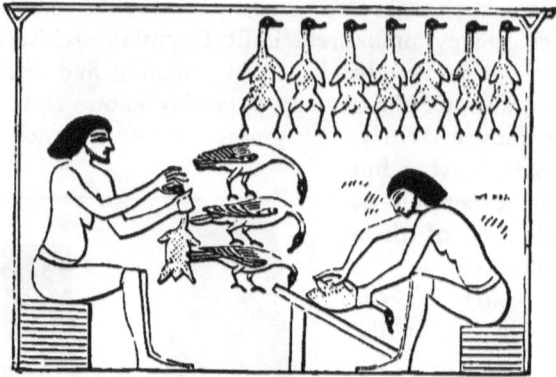

FIG. 4.—POULTERER'S SHOP.

rock; they are ornamented with sculptures and pictures which are full of interest; it has been said that these tombs were built by the Pharaoh, or king, of Joseph's time, and one of the paintings is often spoken of as being a represen-tation of the brethren of Joseph; but of this there is no

FIG. 5.—ROCK-CUT TOMB, BENI-HASSAN.

proof. The colors of the pictures are fresh and bright, and they show that many of the customs and amuse-ments of that long, long ago were similar to our own, and in some cases quite the same. The manufactures of glass and linen, cabinet work, gold ornaments, and other artistic objects are pictured there; the games of ball, draughts, and *morra* are shown, while the animals, birds, and fishes of Egypt are all accurately depicted.

An interesting thing to notice about these tombs is the

way in which the epistyle—the part resting upon the columns—imitates squarely-hewn joists, as if the roof were of wood supported by a row of timbers. When we come to the architecture of Greece we shall see that its most important style, the Doric, arose from the imitation in stone of the details of a wooden roof, and from a likeness between these tombs and the Doric order, this style has been named the Proto-Doric.

The tombs near Thebes which are called the "Tombs of the Kings," and many other Egyptian tombs, are very interesting, and within a short time some which had not before been observed have been opened, and proved to be rich in decorations, and also to contain valuable ornaments and works of art, as well as papyri, or records of historical value.

The most magnificent of all the Egyptian tombs is that of King Seti I., who began to reign in 1366 B.C. He was fond of splendid buildings, and all the architects of his time were very busy in carrying out his plans. His tomb was not discovered until 1817, and was then found by an Italian traveller, whose name, Belzoni, has been given to the tomb. The staircase by which it is entered is twenty-four feet long, and opens into a spacious passage, the walls of which are beautifully ornamented with sculptures and paintings. This is succeeded by other staircases, fine halls, and corridors, all of which extend four hundred and five feet into the mountain in which the tomb is excavated, making also a gradual descent of ninety feet from its entrance. It is a wonderful monument to the skill and taste of the architects who lived and labored more than three thousand years ago.

The two principal cities of ancient Egypt were Memphis and Thebes. The first has been almost literally taken to pieces and carried away, for as other more modern cities have been built up near it, the materials which were first used in the old temples and palaces have been carried here and there, and again utilized in erecting new edifices.

Thebes, on the contrary, has stood alone during all the centuries that have passed since its decline, and there is now no better spot in which to study the ancient Egyptian architecture, because its temples are still so complete that a good idea can be formed from them of what they must have been when they were perfect.   The ruins at Thebes are on both banks of the Nile, and no description can do justice to their grandeur, or give a full estimate of their wonders ; but I shall try to tell something of the palace-temple of Karnak, which has been called " the noblest effort of archi- tectural magnificence ever produced by the hand of man."

The word palace-temple has a strange sound to us because we do not now associate the ideas which the two words represent.   Many palaces of more modern countries and times have their chapels, but the union of a grand temple and a grand palace is extremely rare, to say the least.   Perhaps the Vatican and St. Peter's at Rome repre- sent the idea and spirit of the Egyptian palace-temples as nearly as any buildings that are now in existence.

The Egyptian religion controlled all the affairs of the nation.   The Pharaoh, or king, was the chief of the re- ligion, as well as of the State.   When a king came to the throne he became a priest also, by being made a member of a priestly order.   He was instructed in sacred learning ; he regulated the service of the temple ; on great occasions he offered the sacrifices himself, and, in fact, he was considered not only as a descendant of gods, but as a veritable god. In some sculptures and paintings the gods are represented as attending upon the kings, and after the death of a king the same sort of veneration was paid to him as that given to the gods.   This explains the building of the palace and temple together, and shows the reason why the gods and the kings, and the affairs of religion and of government, could not be separated.   As we study the arts of different coun- tries we are constantly reminded that the religion of a people is the central point from which the arts spring forth. From its teachings they take their tone, and adapt their forms and uses to its requirements. I refer to this fact from time to time because it is important to remember that it underlies much of the art of the world.

It may be said that all the art of Egypt was devoted to the service of its religion. Of course this is true of that used in the decoration of the temples ; it is also true of all that did honor to the kings, because they were regarded as sacred persons, and all their wars and wonderful acts which are represented in sculpture and painting, and by statues and obelisks, are considered as deeds that were performed for the sake of the gods and by their aid.

It was also the religious belief in the immortality of the soul that led the Egyptians to build their tombs with such care, and to provide such splendid places in which to lay the body, which was the house of the spirit.

In the study of Architecture it will also be noted that a country which has no national religion—or one in which the government and the religion have no connection with each other—has no absolutely national architecture. It will have certain features which depend upon the climate, the building materials at command, and upon the general customs of the people ; but here and there will be seen specimens of all existing orders of architecture, and buildings in some degree representing the art of all countries and periods ; such architecture is known by the term composite, because it is composed of portions of several different orders, and has no absolutely distinct character.

This palace-temple of Karnak is made up of a collection of courts and halls, and it is very difficult to comprehend the size of all these parts which go to make up the enormous whole. The entire space devoted to it is almost twice as large as the whole area of St. Peter's at Rome, and four times as great as any of the other cathedrals of

Europe ; a dozen of the largest American churches could be placed within its limits and there still be room for a few chapels. All this enormous space is not covered by roofs, for there were many courts and passages which were always open to the sky, and one portion was added after another, and by one sovereign and another, until the completion of the whole was made long after the Pharaoh who commenced it had been laid in one of the tombs of the kings.

FIG. 6.—THE HALL OF COLUMNS AT KARNAK.

The most remarkable apartment of all is called the great Hypostyle Hall, which high-sounding name means simply a hall with pillars (Fig. 6). This hall and its two pylons, or entrances, cover more space than the great cathedral of Cologne, which is one of the largest and most famous churches of all Europe.

This splendid hall had originally one hundred and thirty-four magnificent columns, of which more than one hundred still remain ; they are of colossal size, some of them being sixty feet high without the base or capital, which would increase them to ninety feet, and their diameter is twelve

feet. This large number of columns was necessary to
uphold the roof, as the Egyptians knew nothing of the
arch, and had no way of supporting a covering over a space
wider than it was possible to cover by beams. The hall
was lighted by making the columns down the middle half
as high again as the others, so that
the roof was lifted, and the light
came in at the sides, which were
left open.

As I must speak often of columns, it is well to say here that
the column or pillar usually consists of three parts—the base, the
shaft, and the capital (Fig. 7). The
base is the lowest part on which the
shaft rests. Sometimes, as in the
Grecian Doric order, the base is
left out. The capital is the head
of the column, and is usually the
most ornamental part, giving the
most noticeable characteristics of
the different kinds of pillars. The
shaft is the body of the pillar, between the base and capital, or all
below the capital when the base is
omitted.

The Egyptian pillars seem to
have grown out of the square stone
piers which at first were used for
support. The square corners were

FIG. 7.—PILLAR FROM
THEBES.
Showing the three parts.

first cut off, making an eight-sided pier ; then some architect carried the cutting farther, and by slicing off each corner once more gave the pillar sixteen sides. The advantage
of the octagonal piers over the square ones was that the
cutting off of sharp corners made it easier for people to

move about between them, while the play of light on the sides was more varied and pleasant to the eye.   The six-

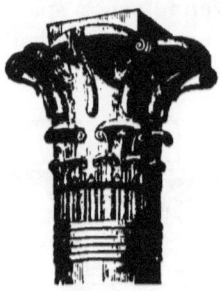

FIG. 8.—SCULPTURED CAPITAL.

teen-sided pillar did not much increase the first of these advantages, while the face of its sides became so narrow that the variety of light and shade was less distinct and attractive.   It is probable that the channelling of the sides of the shaft was first done to overcome this difficulty, by making the shadows deeper and the lights more striking ; and we then have a shaft very like that of the Grecian Doric shown in the picture in Fig. 40, or the Assyrian pillars in Figs. 29 and

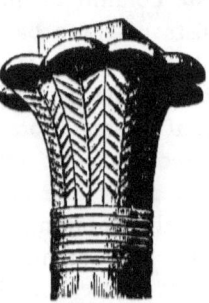

FIG. 9.—PALM CAPITAL.

FIG. 10.— PILLAR FROM SEDINGA.

30.   In the Egyptian pillars it was usual to leave one side unchannelled and ornament it with hieroglyphics.   In time the forms of the Egyptian pillars became very varied, and the richest ornaments were used upon them.   The columns in the hall at Karnak are very much decorated with painting and sculptures, as Fig. 6 shows.   The capitals represent the full-blown flowers and the buds of the sacred lotus, or water-lily.   In other cases the pillars were made to represent bundles of the papyrus plant, and the capitals were often beautifully carved with palm leaves or ornamented with a female head.   (See Figs. 8, 9, and 10).

The whole impression of grandeur made by the Temple of Karnak was increased by the fact that the Temple of

Luxor, which is not far away, is also very impressive and beautiful, and was formerly connected with Karnak by an avenue bordered on each side with a row of sphinxes cut out of stone. These were a kind of statue which belonged to Egyptian art, and originated in an Egyptian idea, although a resemblance to it exists in the art of other ancient countries (Fig. 11).

FIG. 11.—THE GREAT SPHINX.

Before the Temple of Luxor stood Colossi, or enormous statues, of Rameses the Great, who built the temple, and not far distant were two fine obelisks, one of which is now in Paris.

There was much irregularity in the lines and plan of Egyptian palaces and temples. It often happens that the

side walls of an apartment or court-yard are not at right
angles ; the pillars were placed so irregularly and the decorations so little governed by any rule in their arrangement,
that it seems as if the Egyptians were intentionally regard-
less of symmetry and regularity.

The whole effect of the ancient Thebes can scarcely be
imagined ; its grandeur was much increased by the fact that
its splendid buildings were on both banks of the Nile,
which river flowed slowly and majestically by, as if it borrowed a sort of dignity from the splendid piles which it
reflected, and which those who sailed upon its bosom
regarded with awe and admiration.   There are many other
places on the Nile where one sees wonderful ruins of ancient
edifices, but we have not space to describe or even to name
them, and Thebes is the most remarkable of all.

" Thebes, hearing still the Memnon's mystic tones,
Where Egypt's earliest monarchs reared their thrones,
Favored of Jove ! the hundred-gated queen,
Though fallen, grand ; though desolate, serene ;
The blood with awe runs coldly through our veins
As we approach her far-spread, vast remains.
Forests of pillars crown old Nilus' side,
Obelisks to heaven high lift their sculptured pride ;
Rows of dark sphinxes, sweeping far away,
Lead to proud fanes and tombs august as they.
Colossal chiefs in granite sit around,
As wrapped in thought, or sunk in grief profound.

" The mighty columns ranged in long array,
The statues fresh as chiselled yesterday,
We scarce can think two thousand years have flown
Since in proud Thebes a Pharaoh's grandeur shone,
But in yon marble court or sphinx-lined street,
Some moving pageant half expect to meet,
See great Sesostris, come from distant war,
Kings linked in chains to drag his ivory car ;
Or view that bright procession sweeping on,
To meet at Memphis far-famed Solomon,
When, borne by Love, he crossed the Syrian wild,
To wed the Pharaoh's blooming child."

The obelisks of ancient Egypt have a present interest which is almost personal to everybody, since so many of them have been taken away from the banks of the Nile and so placed that they now overlook the Bosphorus, the Tiber, the Seine, the Thames, and our own Hudson River; in

FIG. 12.—CLEOPATRA'S NEEDLES.

truth, there are twelve obelisks in Rome, which is a larger number than are now standing in all Egypt.

The above cut (Fig. 12) shows the two obelisks known as Cleopatra's Needles, as they were seen for a long time at Alexandria. They have both crossed the seas; one was presented to the British nation by Mehemet Ali, and the other, which now stands in Central Park, was a gift to America from the late Khedive of Egypt, Ismail Pasha.

The obelisks were usually erected by the kings to express their worship of the gods, and stood before the temple bearing dedications of the house to its particular deity ; they were covered with the quaint, curious devices which served as letters to the Egyptians, which we call hieroglyphics, and each sovereign thus recorded his praises, and declared his respect for the special gods whom he wished to honor. They were very striking objects, and must have made a fine effect when the temples and statues and avenues of sphinxes, and all the ancient grandeur of the Egyptians was at its height ; and these grave stone watchmen looked down upon triumphal processions and gorgeous ceremonials, and kings and queens with their trains of courtiers passed near them on their way to and from the temple-palaces.

It is always interesting to study the houses and homes of a people—domestic architecture, as it is called ; but one cannot do that in Egypt. It may almost be said that but one ancient home exists, and as that probably belonged to some royal person, we cannot learn from it how the people lived. There were many very rich Egyptians outside of the royal families, and they dwelt in splendor and luxury ; on the other hand, there were multitudes of slaves and very poor people, who had barely enough to eat to keep them alive and enable them to do the work which was set them by their task-masters.

The house of which we speak is at Medinet Habou, on the opposite side of the Nile from Karnak (Fig. 13). It has three floors, with three rooms on each floor, and is very irregular in form. But if we have no ancient houses to study in Egypt, we can learn much about them from the paintings which still exist, and we may believe that the cities which surrounded the old temples fully displayed the wealth and taste of the inhabitants. These pictures show the houses in the midst of gardens laid out with arbors, pavilions, artificial lakes, and many beautiful objects, such as we see in the fine gardens of our own day.

FIG. 13.—PAVILION AT MEDINET HABOU.

After about 1200 B.C. there was a long period of decline in the architecture of Egypt ; occasionally some sovereign tried to do as the older kings had done, but no real revival of the arts occurred until the rule of the Ptolemies was established ; this was after 332 B.C., when Alexander the Great conquered the Persians, who had ruled in Egypt about one hundred and ninety-five years.

Under the Ptolemies Egypt was as prosperous as she had been under the Pharaohs, but the arts of this later time never reached such purity and greatness as was shown in the best days of Thebes ; the buildings were rich and splendid instead of noble and grand, or, as we might say, " more for show" than was the older style.

It is singular that, though the Egypt of the Ptolemies was under Greek and Roman influence, it still remained essentially Egyptian. It seems as if the country had a sort of converting effect upon the strangers who planned and built the temples of Denderah, and Edfou, and beautiful Philæ, and made them try to work and build as if they were the sons of the pure old Egyptians instead of foreign conquerors. So true is this that before A.D. 1799, when scholars began to read hieroglyphics, the learned men of Europe who studied art believed that these later temples were older than those of Thebes.

Outside of Thebes there is no building now to be seen in Egypt which gives so charming an impression of what Egypt might be as does the lovely temple on the island of

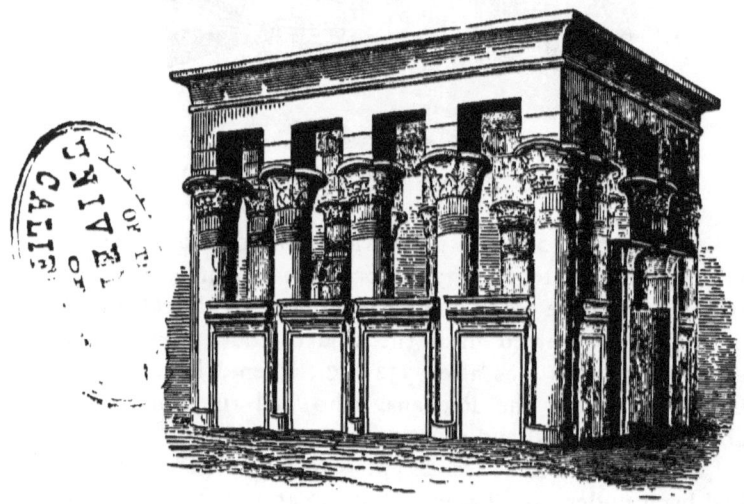

FIG. 14.—TEMPLE ON THE ISLAND OF PHILÆ.

Philæ (Fig. 14). Others are more sublime and imposing, but none are so varied and beautiful.

There is no more attractive spot in Egypt than this island, and when we know that the priests who served in the Temple of Isis here were never allowed to leave the island, we do not feel as if that was a misfortune to them. It was a pity, however, that none but priests were allowed to go there, and in passing I wish to note the fact that this was the most ancient monastery of which we know ; for that it was in simple fact, and the monks lived lives of strict devotion and suffered severe penance.

The buildings at Philæ, as well as most of those of the Ptolemaic age, had the same irregularity of form of which we have spoken before ; their design, as a whole, was fine, but the details were inferior, and it often happens that the sculpture and painting which in the earlier times improved and beautified everything, lost their effect and really injured the appearance of the whole structure.

At first thought one would expect to be able to learn much more about the manners and customs of the later than of the earlier days of Egypt, and to find out just how they arranged their dwellings. But this is not so, for history tells us of nothing save the superstitious religious worship of the conquerors of Egypt. There are no pictures of the houses, or of the occupations and amusements of the people ; no warlike stories are told ; we have no tombs with their instructive inscriptions ; not even the agricultural and mechanical arts are represented in the ruins of this time. The fine arts, the early religion, the spirit of independence and conquest had all died out ; in truth, the wonderful civilization of the days of the pyramid-builders and their descendants was gone, and when Constantine came into power Egypt had lost her place among the nations of the earth, and her grandeur was as a tale that is told.

The weakness of Egyptian architecture lay in its monotony or sameness. Not only did it not develop historically, remaining very much the same as long as it lasted, but the same forms are repeated until, even with all their grandeur, they become wearisome. The plan of the temples varies little ; the tendency toward the shape of the pyramid appears everywhere ; while the powerful influence of the ritual of the Egyptian religion gives a strong likeness among all the places of worship. The Greeks performed upon the ground   If this course be not adopted, serious and expensive mistakes are almost sure to be made, and money wasted in needless alterations.   If you do not know what you want, you are not prepared to build, and should wait till your necessities and tastes have assumed definite forms.   While your house is yet only a paper cottage or villa it may easily be changed to meet your changing whims; but when your thought has once shaped itself in brick and mortar, it has become a matter of enduring record.   See to it that it be such a record as you are willing should be read by posterity.

Adopt no plan hastily, whether conceived by yourself or offered by another.   It should be carefully studied, examined in every light, looked at from every point of view.   There are many things to be taken into consideration.

1. In the first place, your house must be adapted to the site you have chosen.   A plan may be admirable in itself, and yet unsuited to a particular spot.   It must be looked at, then, in reference to the ground it is to occupy; or if the plan be adopted first, the site must be selected in accordance with it.   Not merely the style and general character of a house are influenced by the contour and aspect of the features of the landscape around, but its outlines upon the ground, its arrangement in masses, is equally subject to the great law of fitness.*

2. If one's pecuniary resources are limited, the amount of money which he can appropriate to building will greatly influence the character of his plan.   Reception-rooms, drawing-rooms, libraries, boudoirs, and so on, are certainly desirable; but if you have but seven or eight hundred dollars to expend in building, it would be folly to put them all into your plan. You must be content with a small number of rooms, making, if necessary, several of them serve two or three distinct uses.

Consider first what accommodations are absolutely essential to your comfort, and then what appliances of convenience or luxury you can add.   Do not plan too largely.   Depend upon

* Gervase Wheeler.

it, you will enjoy a much larger sum of happiness in a small house wholly paid for, than in a large one which has involved you in debt.

3. Having decided what sort of a house is best adapted to your site, and what amount of accommodations the sum you purpose to appropriate will secure, consider next how you can make that amount of accommodation best subserve the particular wants and tastes of yourself and family. No two households are exactly alike in their domestic habits, and a house which your neighbor Brown finds "just the thing," would require considerable modification probably to adapt it to your purpose; so in making a plan, or in studying those which we offer in this work, with a view to the adoption of one of them, keep the requirements of your particular household constantly in view, and adopt, modify, or reject accordingly, remembering that the first grand requirement of every dwelling-house is *fitness* or adaptation to its uses.

The fact that individual wants and tastes are infinitely varied, renders it impossible for us to give either directions or plans that will exactly suit individual cases; but we will here briefly advert to some general principles which should govern in the development or choice of a plan.

1. *General Form.*—The largest space in proportion to the extent of the wall may be included in the circular form, but, although round houses have been built, as we shall show further on, this shape is not a desirable one. The octagon approaches the circle in shape and in economy of outside wall. This form is, in our view, open to serious objections, but to give our readers an opportunity to judge for themselves in reference to its advantages and disadvantages, we give plans of octagon houses in another chapter.

O. S. Fowler, in his "Home for All," has advocated this form with an earnestness which could only come from thorough conviction of its superiority over all others. To that work we may have been used by soldiers or guards. The two outer gates were ornamented by sculptured figures of colossal bulls with human heads and other strange designs ; but the inner gates had a plain finish of alabaster slabs. It is thought that arches covered these gateways like some representations of gates which are seen on Assyrian bas-reliefs. Within the gates there is a pavement of large slabs,

FIG. 16.—ENTRANCE TO SMALLER TEMPLE, NIMRUD

in which the marks worn by chariot wheels are still plainly seen.

We learn that the Assyrians made their religion a prominent part of their lives. The inscriptions of the kings begin and end with praises and prayers to their gods, and on all occasions religious worship is spoken of as a principal duty. We know that the monarchs devoted much care to the temples, and built new ones continually ; but it also appears from the excavations that have been made that they devoted the best of their art and the greatest sum of their riches to the palaces of their kings. The temple was far less splendid than the palace to which it was attached as a sort of appendage. This was undoubtedly due to the fact that the Assyrian kings received more than the monarchs of any other ancient people divine honors while still living ; so that the palace was regarded as the actual dwelling of a god. The inner ornamentation of the temples was confined to religious subjects represented on sculptured slabs upon the walls, but no large proportion of the wall was decorated, and the rest was merely plastered and painted in set figures. The gateways and entrances were guarded by sacred figures of colossal bulls, or lions (Fig. 16), and covered with inscriptions ; there was a similarity between the palace entrances and those of the temples.

FIG. 17.—PAVEMENT SLAB FROM KOYUNJIK.

The palaces were always built on artificial platforms, which were made of solid brick or stone, or else the outside walls of the platforms were built of these substances and the middle part filled in with dirt and rubbish. Sometimes the platforms, which were from twenty to thirty feet high, were in terraces and flights of steps led up and down from one to another. It also happened that more than one palace was erected on the same platform ; thus the size and form of the platforms was much varied, and when palaces were enlarged the platforms were changed also, and their shape was often very irregular. The tops of the platforms were paved with stone slabs or bricks, the last being sometimes as much as two feet square ; the pavements were frequently ornamented with artistic designs (Fig. 17), and inscriptions are also found upon them.

At the lower part of the platform there was a terrace on which several small buildings were usually placed, and near by was an important gateway, or, more properly, a propylæum, through which every one must pass who entered the palace from the city. The next cut (Fig. 18) shows one of these grand entrances decorated with the human-headed bulls and the figure of what is believed to be the Assyrian Hercules, who is most frequently represented in the act of

FIG. 18.—REMAINS OF PROPYLÆUM, OR OUTER GATEWAY, KHORSABAD.

strangling a lion. Much rich ornament was lavished on these portals, and the entrance space was probably protected by an arch.

Below these portals, quite down on a level with the city, there were outer gateways, through which one entered a court in front of the ascent to the lower terrace.

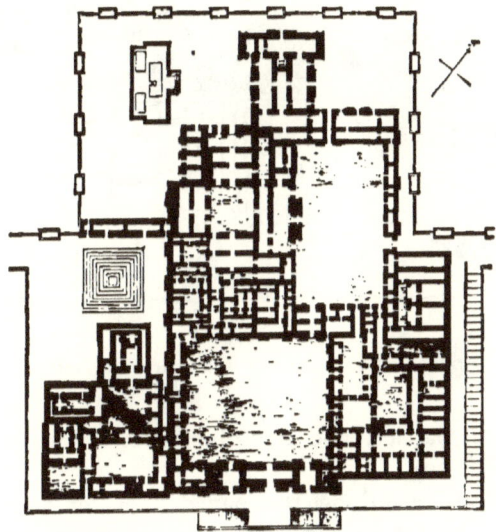

FIG. 19.—PLAN OF PALACE, KHORSABAD.

The principal apartments of the palaces were the courts, the grand halls, and the small, private chambers. The fine palaces had several courts each ; they varied from one hundred and twenty by ninety feet, to two hundred and fifty by one hundred and fifty feet in size, and were paved in the same way as the platforms outside (Fig. 19).

The grand halls were the finest portions of these splendid edifices ; here was the richest ornament, and the walls were lined with sculptured slabs, while colossal bulls, winged genii, and other figures were placed at the entrances. Upon the slabs the principal events in the lives of the monarchs were represented, as well as their portraits, and religious

FIG. 20.—RELIEF FROM KHORSABAD. A TEMPLE.

ceremonies, battles, and many incidents of interest to the nation (Fig. 20).

The slabs rested on the paved floors of the halls and reached a height of ten or twelve feet ; above them the walls were of burnt brick, sometimes in brilliant colors ; the whole height of the walls was from fifteen to twenty feet. The smaller chambers surrounded these grand halls, and the number of rooms was very large ; in one palace which has been but partially explored there are sixty-eight apartments, and it is not probable that any Assyrian palace had less than forty or fifty rooms on its ground floor. Of all the palaces which have been examined that of Khorsabad is best known and can be most exactly described. It is believed that Sargon, a son of Sennacherib, built it, and it is very splendid.

After entering at the great portal one passes through various courts and corridors ; these are all adorned with sculptures such as have been described above ; at length one reaches the great inner court of the palace, which was a square of about one hundred and fifty feet in size. This court had buildings on two sides, and the other sides extended to the edge of the terrace of the platform on which the palace was built, and commanded broad views of the open country. On one side the buildings contained the less important apartments of the officers of the court; the grand state apartments were on the other side. There were ten of these at Khorsabad; five were large halls, four were smaller chambers, and one a long and narrow room. Three of the large halls were connected with one another, and their decorations were by far the most splendid of any in the palace. In one of them the sculptures represented the king superintending the reception and chastisement of prisoners, and is called the " Hall of Punishment." The middle hall has no distinguishing feature, but the third opened into the " Temple Court," on one side of which the small temple was situated. The lower sculptures of the middle and third halls represented the military history of Sargon, who is seen in all sorts of soldier-like positions and occupations; some of the upper sculptures represent religious ceremonies.

On one side of the Temple Court there were several chambers called Priests' Rooms, but the temple itself and the portions of the palace connected with it are not as well preserved as the other parts, and have nothing about them to interest us in their study.

The palaces of Nineveh are much less perfect than the palace-temples of Thebes, and cannot be described with as much exactness. There is no wall of them still standing more than sixteen feet above the ground, and we do not even know whether they had upper stories or not, or how they were lighted—in a word, nothing is positively known about them above the ground floors, and it is very strange that the sculptures nowhere represent a royal residence. But what we do know of the Assyrians proves that they equalled and perhaps excelled all other Oriental nations as architects and designers, as well as in other departments of art and industry.

FIG. 21.—RESTORATION OF AN ASSYRIAN PALACE.

This representation of an Assyrian palace (Fig. 21) is a restoration, as it is called, being made up by a careful study of the remains and such facts as can be learned from bas-reliefs, and cannot be wholly unlike the dwellings of the king-gods. It is pleasing in general appearance, and for lightness and elegance is even to be preferred to Egyptian architecture, though it is far inferior in dignity and impressiveness.

The Assyrians knew the use of both column and arch, but never developed either to any extent. They also employed the obelisk, and it is noticeable that instead of terminating it with a pyramid, as was the case in Egypt, they capped it with the diminishing terraces, which is the fundamental form which underlies all the architecture of the country, as the smooth pyramid is the most prominent element in the architecture of Egypt.

BABYLON.

It is probable that Babylon was the largest and finest of all the ancient cities. The walls which surrounded it, together with its hanging gardens, were reckoned among the " seven wonders of the world " by the ancients. Its walls were pierced by a hundred gates and surmounted by two hundred and fifty towers ; these towers added much to the grand appearance of the city ; they were not very high above the walls, and were probably used as guard-rooms by soldiers.

The River Euphrates ran through the city. Brick walls were built upon its banks, and every street which led to the river had a gateway in these walls which opened to a sloping landing which extended down to the water's edge ; boats were kept at these landings for those who wished to cross the stream. There was also a foot-bridge across the river that could be used only by day, and one writer, Diodorus, declares that a tunnel also existed which joined the two sides of the river, and was fifteen feet wide and twelve feet high in the inside.

The accounts of the " Hanging Gardens" make it seem that they resembled an artificial terraced mountain built upon arches of masonry and covered with earth, in which grew trees, shrubs, and flowers. It is said by some writers that this mountain was at least seventy-five feet high, and occupied a square of four acres ; others say that in its highest part it reached three hundred feet ; but all agree that it was a wonderful work and very beautiful.

In the interior of the structure machinery was concealed

which raised water from the Euphrates and filled a reservoir at the summit, from which it was taken to moisten the earth and nourish the plants. Flights of steps led up to the top, and on the way there were entrances to fine apartments where one could rest. These rooms, built in the walls which supported the structure, were cool and pleasant, and afforded fine views of the city and its surroundings. The whole effect of the gardens when seen from a distance was that of a wooded pyramid. It seems a pity that it should have been called a " Hanging Garden," since, when one knows how it was built, this name is strangely unsuitable, and carries a certain disappointment with it.

The accounts of the origin of this garden are interesting. One of them says that it was made by Semiramis, a queen who was famous for her prowess as a warrior, for having conquered some cities and built others, for having dammed up the River Euphrates, and performed many marvellous and heroic deeds. It is not probable that any woman ever did all the wonders which are attributed to Semiramis, but we love to read these tales of the old, old time, and it is important for us to know them since they are often referred to in books and in conversation.

Another account relates that the gardens were made by Nebuchadnezzar to please his Median queen, Amytis, because the country round about Babylon seemed so barren and desolate to her, and she longed for the lovely scenery of her native land.

What we have said will show that the Babylonians were advanced in the science of such works as come more properly under the head of engineering ; their palaces were also fine, and their dwelling-houses lofty ; they had three or four stories, and were covered by vaulted roofs. But the Babylonians, like the Egyptians, lavished their best art upon their temples. The temple was built in the most prominent position and magnificently adorned. It was

usually within a walled inclosure, and the most important
temple at Babylon, called that of Belus, is said to have had
an area of thirty acres devoted to it.   The chief distinguishing feature of a Babylonish temple was a tower built in
stages (Fig. 22).

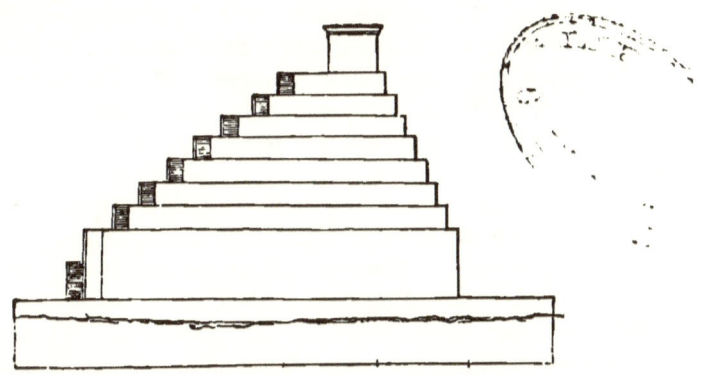

Fig. 22.—Elevation of the Temple of the Seven Spheres at Borsippa.

The number of the stages varied, eight being the largest.
At the summit of the tower there was a chapel or an altar,
and the ascent was by steps or an inclined plane which
wound around the sides of the tower.   The Babylonians
were famous astronomers, and it is believed that these
towers were used as observatories as well as for places of
worship.   At the base of the tower there was a chapel for
the use of those who could not ascend the height, and near
by, in the open air, different altars were placed, for the
worship of the Babylonians included the offering of sacrifices.

Very ancient writers describe the riches of the shrines at
Babylon as being of a value beyond our belief.   They tell
of colossal images of the gods of solid gold ; of enormous
lions in the same precious metal ; of serpents of silver,
each of thirty talents' weight (a talent equalled about two

thousand dollars of our money), and of golden tables, bowls, and drinking-cups, besides magnificent offerings of many kinds which faithful worshippers had devoted to the gods. These great treasures fell into the hands of the Persians when they conquered Babylon.

The Birs-i-Nimrud has been more fully examined than any other Babylonish ruin, and a description of it can be given with a good degree of correctness. As it now stands, every brick in it bears the name of Nebuchadnezzar ; it is believed that he repaired or rebuilt it, but there is no reason to think that he changed its plan. Be this as it may, it is a very interesting ruin (Fig. 23). It was a temple raised on a platform and built in seven stages ; these stages represented the seven spheres in which the seven planets moved (according to the ancient astronomy), and a particular color was assigned to each planet, and the stages colored according to this idea. That of the sun was golden ; the moon, silver ; Saturn, black ; Jupiter, orange ; Mars, red ; Venus, pale yellow, and Mercury, deep blue.

It is curious to know how the various colors were obtained. The lower stage, representing Saturn, was covered with bitumen ; that of Jupiter was faced with bricks burned to an orange color ; that of Mars was made of bricks from a bright red clay and half burned, so that they had a blood-red tint ; the stage dedicated to the sun was probably covered with thin plates of gold ; that of Venus had pale yellow bricks ; that of Mercury was subjected to intense heat after it was erected, and this produced vitrification and gave it a blue color ; and the stage of the moon was coated in shining white metals.

Thus the tower rose up, all glowing in colors and tints as cunningly arranged as if produced by Nature herself. The silvery, shining band was probably the highest, and had the effect of mingling with the bright sky above. We can scarcely understand how glorious the effect must have been, and when we try to imagine it, and then think of the present wretched condition of these ruins, it gives great force to the prophecies concerning Babylon which foretold that her broad walls should be utterly broken down, her gates burned with fire, and the golden city swept with the besom of destruction.

FIG. 23.—BIRS-I-NIMRUD, NEAR BABYLON.

We know so little of the arrangement of the palaces of Babylon that we cannot speak of them in detail. They differed from those of Assyria in two important points: they are of burnt bricks instead of those dried in the sun which the Assyrians used, and at Babylon in the decoration of the walls colored pictures upon the brick-work took the place of the alabaster bas-reliefs which were found in the palaces of Nineveh.

These paintings represented hunting scenes, battles, and other important events, and were alternated with portions of the wall upon which were inscriptions painted in white on a blue ground, or spaces with a regular pattern of rosettes or some fixed design in geometrical figures. A sufficient number of these decorations have been found in the ruins of Babylon to prove beyond a doubt that this was the customary finish of the walls. We also know that the houses of Babylon were three or four stories in height, but were rudely constructed and indicate an inferior style of domestic architecture.

PERSIA.

The Persians were the pupils of the Assyrians and Baby-lonians in Art, Learning, and Science, and they learned their lessons so well that they built magnificent palaces and tombs. Temples seem to have been unimportant to them, and we know nothing of any Persian temple remains that would attract the attention of travellers or scholars.

The four most important Persian palaces of which we have any good degree of knowledge are that of Ecbatana, the ruins of which are very imperfect ; a second at Susa, of which the arrangement is known ; a third at Persepolis, which is not well enough preserved· for any exact descrip-tion to be given ; and a fourth, the so-called Great Palace, near Persepolis, in which the latest Persian sovereigns lived. This magnificent palace was burned by Alexander the Great before he or his soldiers had seen its splendor. The story is that he made a feast at which Thais, a beautiful and wicked woman, appeared, and by her arts gained such power over Alexander that he consented to her proposal to fire the palace, and the king, wearing a crown of flowers upon his head, seized a torch and himself executed the dreadful deed, while all the company followed him with acclamations, singing, and wild shouts. At last they surrounded and danced about the dreadful conflagration.

The poet Dryden wrote an ode upon "Alexander's Feast" in 1697 which has a world-wide reputation. I quote a few lines from it:

> "'Twas at the royal feast for Persia won
>   By Philip's warlike son :
> Aloft, in awful state,
> The godlike hero sate
>   On his imperial throne ;
> His valiant peers were placed around,
> Their brows with roses and with myrtles bound
> (So should desert in arms be crowned) ;
>   The lovely Thais by his side
>   Sate, like a blooming Eastern bride,
> In flower of youth and beauty's pride.
> Happy, happy, happy pair !
>   None but the brave,
>   None but the brave,
> None but the brave deserves the fair.

> "Behold how they toss their torches on high,
> How they point to the Persian abodes,
> And glittering temples of their hostile gods !
> The princes applaud with a furious joy,
> And the king seized a flambeau with zeal to destroy ;
>   Thais led the way
>   To light him to his prey,
> And, like another Helen, fired another Troy."

Much study and time has been given to the examination of the ruins of Persepolis, and the whole arrangement of the city has been discovered and is made plain to the student of these matters by means of the many charts, plans, and photographs of it which now exist. I shall try to tell you something of the Great Palace of Persepolis, and the other palaces near it and on the platform with it, for the Persians, like the Assyrians and Babylonians, built their palaces upon platforms. This one of which we speak was distinct from the city, but quite near it, and is in almost perfect condition.

It is composed of large masses of hewn stone held together by clamps of iron or lead. Many of the blocks in this platform wall are so large as to make their removal from the quarries and their elevation to the required height a difficult mechanical task, which could only have been performed by skilled laborers with good means for carrying on their work. The wall was not laid in regular blocks, but was like this plate (Fig. 24).

FIG. 24.—MASONRY OF GREAT PLATFORM, PERSEPOLIS.

The platform was not of the same height in all its parts, and seems to have been in several terraces, three of which can still be seen. The buildings were on the upper terrace, which is about forty-five feet above the plain and very large ; it is seven hundred and seventy feet long and four hundred feet wide. The staircases are an important feature of these ruins, and when all the palaces were in perfection these broad steps, with their landings and splendid decorations, must have made a noble and magnificent effect. The ascent of the staircases was so gradual and easy that men went up and down on horseback, and travellers now ascend and descend in this way.

There is little doubt that the staircases of Persepolis were the finest that were ever built in any part of the world, and on some of them ten horsemen could ride abreast. The broadest, or platform staircase, is entirely

without ornament; another which leads from the platform
up to the central or upper terrace is so elaborately decorated that it appears to be covered with sculptures. There
are colossal representations of lions, bulls, Persian guards-
men, rows of trees, and continuous processions of smaller
figures. In some parts the sculptures represent various
nations bringing trib-
utes to the Persian
monarch; in other
parts all the different
officers of the court
and those of the army
are seen, and the
latter appear to be
guarding the stairs.
(See Fig. 25.)

In a conspicuous
position on this or-
namental staircase
there are three slabs;
on two there is no
design of any sort;
on the third an in-
scription says that
this was the work of
" Xerxes, the Great

FIG. 25.—PARAPET WALL OF STAIRCASE,
PERSEPOLIS (RESTORED).

King, the King of Kings, the son of King Darius, the
Achæmenian." This inscription is in the Persian tongue,
and it is probable that it was the intention to repeat it on
the slabs which are left plain in some other languages, so
that it could easily be read by those of different nations;
it was customary with the ancients to repeat inscriptions in
this way.

The other staircases of this great platform are all more
or less decorated with sculptures and resemble that de-

scribed ; they lead to the different palaces, of which there
are three. The palaces are those of Darius, Xerxes, and
Artaxerxes Ochus, and besides these there are two great
pillared halls ; one of these is called the " Hall of One
Hundred Columns," and the other *Chehl Minar*, or the
" Great Hall of Audience."

This view of the palace of Darius gives an idea of the
appearance of all these buildings. A description of them

FIG. 26.—RUINS OF THE PALACE OF DARIUS, PERSEPOLIS.

would be only a wordy repetition of the characteristics of
one apartment and hall after another, and I shall leave
them to speak of the magnificent halls which are the glory
of the ruins of Persepolis, and the wonders of the world to
those who are acquainted with the architectural monuments
of the Turkish, Greek, Roman, Moorish, and Christian
nations. (See Fig. 26.)

The Hall of a Hundred Columns was very splendid, as
one may judge from this picture of its gateway (Fig. 27) ;
but the *Chehl Minar*, or Great Hall of Audience, which is

also called the Hall of Xerxes, was the most remarkable of all these edifices. Its ruins occupy a space of almost three hundred and fifty feet in length and two hundred and forty-six feet in width, and consist principally of four different kinds of columns. One portion of this hall was arranged in a square, in which there were six rows of six pillars each,

FIG. 27.—GATEWAY OF HALL OF A HUNDRED COLUMNS.

and on three sides of this square there were magnificent porches, in each of which there were twelve columns; so that the number of pillars in the square was thirty-six, and that of those in the three porches was the same. These porches stood out boldly from the main building and were grand in their effect.

The columns which remain in various parts of this hall

FIG. 28.—DOUBLE HORNED LION CAPITAL.

are so high that it is thought that they must originally have measured sixty-four feet throughout the whole building.

FIG. 29.—COMPLEX CAPITAL AND BASE OF PILLARS, PERSEPOLIS.

The capitals of the pillars were of three kinds : the double Horned Lion capital (Fig. 28) was used in the eastern porch, and was very simple ; in the western porch was the double Bull capital, which corresponded to the first in size and general form, the difference being only in the shape of the animal.

The north porch faced the great sculptured staircase, and was the real front of the hall. On this side the columns were much ornamented. The following plates show the entire design of them, and it will be seen that the bases were very beautiful (Figs. 29 and 30).

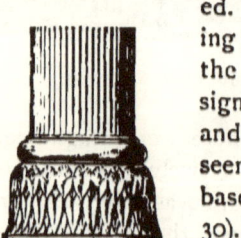

FIG. 30.—BASE OF ANOTHER PILLAR, PERSEPOLIS.

The capitals have three distinct parts ; at the bottom is a sort of bed of lotus leaves, part of which are turned down, and the others standing up form a kind of cup on which the next section above rests. The middle section is fluted and has spiral scrolls or volutes, such as are seen in Ionic capitals, only here they are in a perpendicular position instead of the customary horizontal one. The upper portion had the same double figures of bulls as were on the columns of the western colonnade. The decoration on the bases was made of two or three rows of hanging lotus leaves, some round and others pointed in form. The shafts of these pillars were formed of different blocks of stone joined by iron cramps ; they were cut in exact and regular flutings, numbering from forty-eight to fifty-two on each pillar.

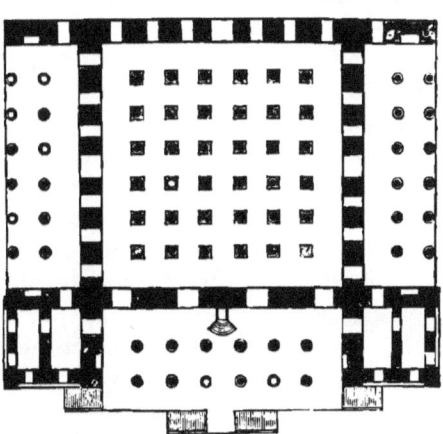

This plan of the Hall of Audience will help you to understand its arrangement more clearly (Fig. 31).

FIG. 31.—GROUND PLAN (RESTORED) OF HALL OF XERXES, PERSEPOLIS.

The square with the thirty-six columns, and the three porches with twelve columns each, are distinctly marked. The most ornamental pillars were on the side with the entrance or gateway. The two small rooms on the ends of the main portico may have been guard-rooms.

We can only regret that, while we know certain things about this hall, there is still much of which we know nothing. However, there are many theories concerning it. Some authorities believe that it was roofed, while others think that it was open and protected only by curtains and hangings, of which the Persians made much use. As we cannot know positively about it, and Persepolis was the spring residence of the Persian kings, it is pleasant to fancy that this splendid pillared hall was a summer throne-room, having beautiful hangings that could be drawn aside at will, admitting all the spicy breezes of that sunny land, and realizing the description of the palace of Shushan in the Book of Esther, which says, " In the court of the garden of the king's palace ; where were white, green, and blue hangings, fastened with cords of fine linen and purple to silver rings and pillars of marble ; the beds were of gold and silver, upon a pavement of red, and blue, and white, and black marble."

Here the king could receive all those who sought him ; the glorious view of the plains of Susa and Persepolis, the breezes which came to him laden with the odors of the choicest flowers would soothe him to content, and realize his full desire for that deep breath from open air which gives a sense of freedom and power. We know that no Oriental, be he monarch or slave, desires to live beneath a roof or within closed doors.

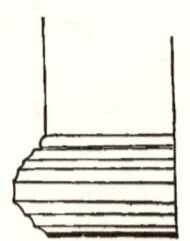

Fig. 32.—Part of a Base of the Time of Cyrus, Pasargadæ.

The column was in Persia developed with a good deal of originality and much artistic feeling ; and one fine base of the time of Cyrus is especially interesting for its close resemblance to the base of certain Ionic pillars afterward made in Greece (Fig. 32).

The tombs of the royal Persians were usually hewn out of the solid rock ; the tomb of Cyrus, only, resembles a little house ; this plate gives a representation of it (Fig. 33).

The one apartment in this tomb is about eleven feet

FIG. 33.—THE TOMB OF CYRUS.

long, seven feet broad, and seven feet high ; it has no window, and a low, narrow doorway in one of the end walls is the only entrance to it. Ancient writers say that the body of Cyrus in a golden coffin was deposited in this tomb.

Seven other tombs have been explored ; they are excavations in the sides of the mountains high enough to be prominent objects to the sight, and yet difficult of approach. The fronts of these tombs are much ornamented, and the internal chambers are large ; there are recesses for the burial-cases, and these vary in number, some having only space for three bodies. The tomb of Darius had three recesses, in each of which there were three burial-cases ; but this was an unusually large number. The tombs near Persepolis are the finest which have yet been examined.

The most noticeable characteristic of Persian architecture is its regularity. The plans used are simple, and only straight lines occur in them ; thus, all the angles are right angles. The columns are regularly placed, and the two sides of an apartment or building correspond to each other. The magnificent staircases, and the abundance of elegant columns which have been called "groves of pillars" by some writers, produced a grand and dignified effect. The huge size of the blocks of stone used by Persian builders gives an impression of great power in those who planned their use, and demands for them the respect of all thoughtful students of these edifices.

The faults of this architecture lay in the narrow doorways, the small number of passages, and the clumsy thickness of the walls. But these faults are insignificant in comparison with its beauties, and it is all the more to be admired that it was invented by the Persians, not copied from other nations, and there is little doubt that the Greeks profited by its study to improve their own style, and through this study substituted lightness and elegance for the clumsy and heavy effect of the earliest Grecian architecture.

JUDEA.

There is so much of religious, historical, romantic, and poetical association with the land of Judea, that it is a disappointment to know that there are no remains of Judean architecture from which to study the early art-history of that country ; it is literally true that nothing remains.

The ruins of Jerusalem, Baalbec, Palmyra, Petra, and places beyond the Jordan are not Jewish, but Roman remains. The most interesting remnant is a passage and gateway which belonged to the great temple at Jerusalem. This passage is situated beneath the platform of the temple ; it is called "The Gateway Huldah." The width of it is forty-one feet, and at one point there is a magnificent pillar, called a monolith, because it is cut from a single stone. This pillar supports four arches, which divide the passage into as many compartments, each one of which has a flat dome. On these domes or roofs there were formerly beautiful ornamental designs, one of which remains, and is like this picture (Fig. 34). Its combination of Oriental and Roman design proves that it cannot be very old, but must have been made after the influence of the Romans had been felt in Judea.

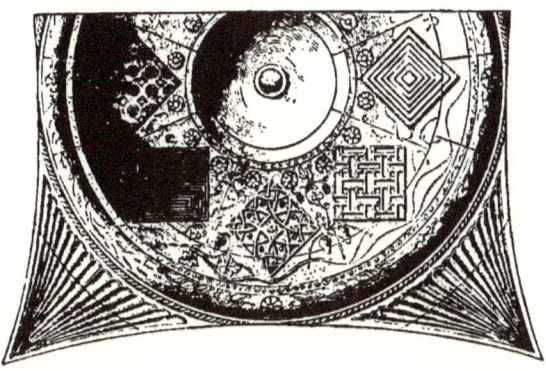

FIG. 34.—ROOF OF ONE OF THE COMPARTMENTS OF THE GATE HULDAH.

Since the excavations in Assyria, and through the use of the knowledge obtained there and in other ancient countries, and by comparing this with the descriptions of the Bible and the works of Josephus, some antiquarians have made plans and drawings of what they believe that the temple at Jerusalem must have been at the time of the Crucifixion. The result of this work has little interest, for two reasons : first, because we do not know that it is correct ; second, because even at the time to which it is ascribed, it was not the ancient temple of Solomon. That had been destroyed, and after the return of the Jews from the Captivity, was rebuilt ; again, it had been changed and restored by the Romans under Herod, so that it had little in reality, or by way of association, to give it the sacred and intense interest

for us which would belong to the true, ancient temple **at**
Jerusalem.

" Lost Salem of the Jews, great sepulchre.
    Of all profane and of all holy things,
  Where Jew and Turk and Gentile yet concur
    To make thee what thou art, thy history brings
    Thoughts mixed of joy and woe.   The whole earth rings
  With the sad truth which He has prophesied,
    Who would have sheltered with his holy wings
  Thee and thy children.   You his power defied ;
  You scourged him while he lived, and mocked him as he died !

" There is a star in the untroubled sky,
    That caught the first light which its Maker made,—
  It led the hymn of other orbs on high ;
    'Twill shine when all the fires of heaven shall fade.
    Pilgrims at Salem's porch, be that your aid !
  For it has kept its watch on Palestine !
    Look to its holy light, nor be dismayed,   ·
  Though broken is each consecrated shrine,
  Though crushed and ruined all which men have called divine."

GREECE.

The earliest history of Greece is lost in what we may
call the Age of Legend.   From that period have come to
us such marvellous stories of gods and goddesses, and all
sorts of wonderful happenings and doings, that even the
most serious and wise scholars can learn little about it, and
it remains to all alike a kind of delightful fairy-land.

Back to that remote age one can send his fancy and
imagination to feast upon the tales of wondrous bravery,
passionate love, dire revenge, and supernatural occurrences
of every sort until he is weary of it all.   Then he is glad to
come back to his actual life, in which cause and effect are
so much more clearly seen, and which, if more matter-of-
fact, is more comfortable than the hap-hazard existence of
those remarkable beings who were liable to be changed into

beasts, or trees, or almost anything else at a moment's notice, or to be whisked away from the midst of their families and friends and set down to starve in some desolate place where there was nothing to eat, and no one to listen to complaints of sorrow or hunger.

This legendary time in Grecian history begins nobody knows when, and ends about one thousand years before the birth of Christ. Our only knowledge of it comes from the mythology which we have inherited from the past, and the two poems of Homer, called the " Iliad " and the " Odyssey."

The "Iliad" recounts the anger of Achilles and all that happened in the Trojan War ; the "Odyssey" relates the wonderful adventures of Ulysses. Probably Homer never thought of such a thing as being an historian—he was a poet—much less did he dream of being the only historian of any certain time or age ; but since, in the course of his poems, he refers to the manners and customs of the years that had preceded him, and gives accounts of certain past events, he is, in truth, the prime source from which we learn the little that we know of the prehistoric days in Greece.

It is believed that Homer wrote about 850 B.C., and after that date we have nothing complete in Greek literature until the time of Herodotus, who is called the " Father of History" and was born in 484 B.C. Thus four centuries between Homer and Herodotus are left with no authoritative writings.

The legendary or first period of Greek history was followed by five hundred years more of which we have no continuous history ; but facts have been gathered here and there from the works of various authors which make it possible to give a reliable account of the Greece of that time. For our purpose in this book we go on to a still later time, or a third period, which began about 500 B.C., in which the

architecture and art which we have in mind, when we use
the general term Greek Art, originated.

It is true that before this temples had been erected of
which we have some knowledge, and the elegant and ornate
articles which Dr. Schliemann has found in his excavations
at Troy and Mycenæ prove that the art of that remote time
reached a high point of excellence. The temples and other
buildings of which we know anything, and which belonged
to the second period, were clumsy and rude when compared
with the perfection of the time which we propose to study.

Before we speak of any one edifice it is best to understand something of the various orders of Greek architecture, more especially as the terms which belong to it and
had their origin in it are now used in speaking of architecture the world over, and from being first applied to Greek
art have grown to be general in their application.

In the most ancient days of Greece the royal fortresses
were the finest structures, but in later days the temple
became the supreme object upon which thought and labor
were lavished. The public buildings which served the uses
of the whole people were second in consideration, while the
private dwellings were of the least importance of all. The

Greek temple was built upon a raised
structure like those of Assyria and other
Oriental nations, but the Greek temple
was much smaller, and by a dignified and
simple elegance in detail, and a harmony
in all its parts, it expressed a more noble
religious sentiment than could be con-
veyed by all the vast piles of massive
confusion that had abounded in more
Eastern lands.

Fig. 36.—Temple of
Diana, Eleusis.

The earliest and simplest Greek temples were merely
small, square chambers made to contain an image of a god,
and in later times, when the temples came to be splendid

Fig. 35.—Gravestone from Mycenæ (Schliemann).

FIG. 37.—SMALL TEMPLE AT RHAMNUS.

and grand, the apartment containing the sacred image was
still called the *cella* or cell, as it had been named from the
first. The simplest form of temple was like the little cut
(Fig. 36), and had two pillars in the centre of the front
and two square pilasters at the front end of the side walls.
These pilasters are called *antæ*, and the whole style of the
building is called *distyle in antis;* the word distyle denotes
the two pillars, and the expression means two pillars with
antæ.

The above picture shows the next advance that was
made in form (Fig. 37). A porch was added to the cell,.

FIG. 39.—THE PARTHENON, Athens. (RESTORED.)

the two parts being separated by a wall with a doorway in it. After a time the number of pillars in front was increased to six, and the two outer ones were the first of a row which extended along the entire length of the sides of the temple, thus forming a peristyle, or a row of columns entirely around the cell ; the cell itself remained, according to the original plan, in the centre of the building. The ground plan of such a temple is given in the next wood-cut (Fig. 38).

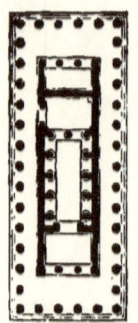

FIG. 38.—PLAN OF TEMPLE OF APOLLO, BASSÆ.

A large proportion of the Greek temples were built in this manner, and were called *hexastyle* from the six columns on the front.

The different orders of ancient Greek architecture are called the Doric, the Ionic, and the Corinthian. The Greeks were very fond of the Doric order, and used it so extensively as to make it almost exclusively their own. The picture of the Parthenon will help you to understand the explanations of the characteristics of the Doric order (Fig. 39).

As you see, the pillars had no base, but rested directly on the upper plinth of the foundation of the building. The shaft of the column is cut in flutings, and the number of them varies from sixteen to twenty ; the latter number being most frequently used. The capital of the column is divided into two portions ; the lower one is called the *echinus*, and projects beyond the shaft and supports a square tile or block which is called the *abacus*, and this is the architectural name for the upper member of all capitals to columns. The *architrave* or principal beam above these columns rests directly on the capitals and runs around the building. This architrave is made of separate blocks of marble or stone, and is finished at the top by a small strip of the same materials, which is called a *tenia*. This cut, which gives a section of the Parthenon on a larger scale than the last picture, will enable you to find the different portions more easily (Fig. 40).

Above the architrave and resting on it is the *frieze ;* this is ornamented with fluted spaces called *triglyphs,* because they are cut in three flutings. The spaces between the

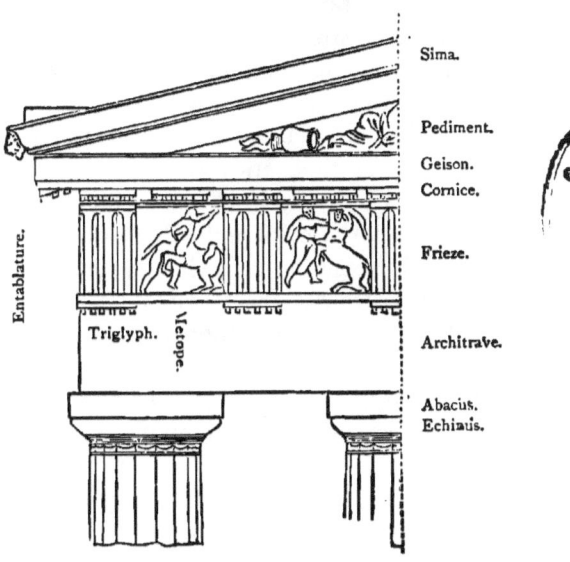

FIG. 40.—FROM THE PARTHENON, ATHENS.

triglyphs are called *metopes,* and sometimes left plain, and sometimes ornamented with sculptures, as is the case in the frieze of the Parthenon. Under the triglyphs six little blocks, or drops, are placed so that they lay over the architrave. Above the frieze there is another narrow strip, or tenia, like that upon the architrave. Above all this rests the *cornice,* and underneath the cornice are one or more rows of the small, drop-like blocks such as make the lower finish of the triglyphs ; in the lower band of the cornice separate blocks are placed over each triglyph and each metope, with a small space between.

It is important to know that the architrave, frieze, and cornice, all taken together, form what is called the *entablature ;* and the entablature occupies the whole of the broad space between the top of the capitals of the pillars and the lower edge of the roof.

The triangular space formed by the sloping of the roof upon the ends of a building is called the *pediment*, and, as you will see in the picture of the Parthenon, its pediment was ornamented with elaborate sculptures which are spoken of in the volume of this series which is devoted to that art. It was customary to thus ornament the pediment and to paint the walls of the cella and other portions of the building, so that while the pure Doric style seems at first sight to be stiff and straight in its effect, it becomes rich and ornamental by the use of sculpture and painting, and yet remains solid and stable.

The Doric style may be regarded as a native growth in Greece, as almost every detail of its construction and its ornaments may be traced back to the early wooden buildings of the people, as the architecture of the tombs of Beni-Hassan had been. The triglyphs, for instance, represent the ends of the beams upon which the rafters rested, while the bas-reliefs between took the place of the votive offerings which in the primitive temples were placed in the open spaces between the beams. It is not necessary here to go into all the particulars of this resemblance, which perhaps learned men have sometimes carried too far, and which are rather difficult to understand ; it is enough to say that there are excellent reasons for regarding the theory as, upon the whole, sound, although, of course, the Grecian architects modified and enriched the forms which the simple timber work had suggested.

The next great order was called the Ionic, and has a

close relation with certain forms found in Asia Minor. This picture of an Ionic capital and entablature is taken from the Temple of Athena at Priene (Fig. 41). Its scroll-like capital recalls those of the pillars in the Great Hall of Xerxes at Persepolis, shown in Figs. 28 and 29, and many examples of even closer resemblance might be given. The order differed from the Doric principally in the ornamentation of its capitals and in the fact that the

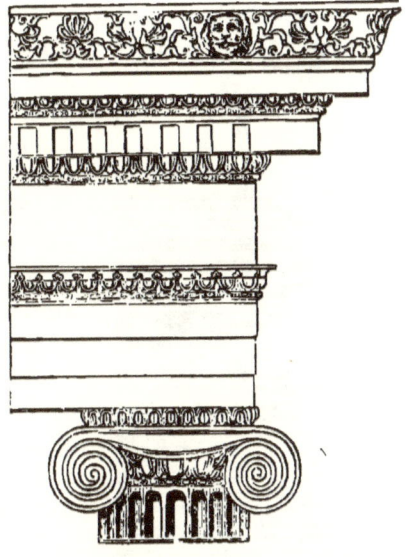

FIG. 41.—IONIC ARCHITECTURE.

columns have bases. These cuts show different kinds of bases belonging to the Ionic order. The first is from the

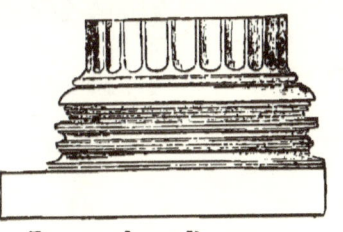

FIG. 42.—IONIC BASE, FROM PRIENE.

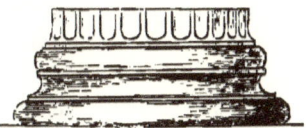

FIG. 43.—ATTIC BASE.

temple at Priene (Fig. 42), and the second is the form known as the Attic base (Fig. 43). The third is especially interesting from its close resemblance to the ancient Persian base shown in Fig. 32, and is another illustration of the Eastern origin of this order (Fig. 44).

The Ionic capital is very easily recognized by its spiral projections, or scrolls, which are called volutes (Fig. 45).

These are so placed that they present a flat surface on the opposite sides of the capital, like the picture below (Fig. 46) ; sometimes the volutes are finished by a rosette in the centre.

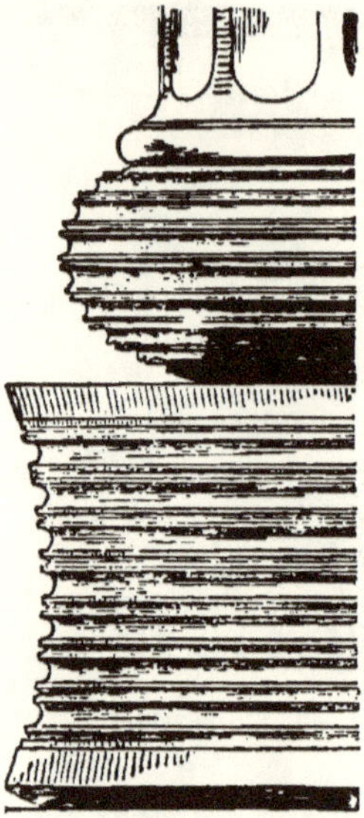

The shaft of the Ionic column is sometimes plain and sometimes fluted ; the flutings number twenty-four, and are separated by a narrow, plain band or fillet. In some ancient examples of the Ionic order the entire entablature is left plain, but in many instances there are bands of carvings, as in the first Ionic example given above ; in some modern Italian architecture even more ornament has been added.

The three, or sometimes two, layers or bands of stone which form the Ionic architrave project a little, each one more than the other, and the ornamented band above it serves to separate it from the frieze so as to make

FIG. 44.—BASE FROM TEMPLE OF HERA, SAMOS.

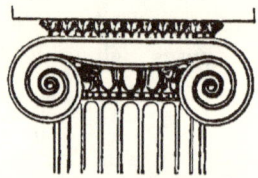

FIG. 45.—IONIC CAPITAL (FRONT VIEW).

FIG. 46.—IONIC CAPITAL (SIDE VIEW).

these two portions of the entablature quite distinct from each other. The frieze is never divided into set spaces as in the Doric order, but when ornamented has a continuous design in relief.

The lower part of the cornice is frequently cut in little pieces or dentals which form what is called the " tooth-like

ornament ;" these have the effect of hanging from underneath the cornice. There is a certain pleasing effect in
Ionic architecture which, perhaps, appeals to our taste at
first sight more forcibly than does the severe elegance of
the Doric order. Nevertheless, the latter is a higher type
of art, and it is not probable that it can ever be superseded
by any new invention or
lose the prestige which
it has held so long.

That which is called
the Corinthian order
differs very little from
the Ionic except in the
capital, but as this was
so prominent a member
of the Ionic style, the
difference seems greater
than it really is. It is
therefore not necessary
to speak of its parts in
detail. The Choragic
Monument of Lysicrates
at Athens is as good a
specimen of the order as
remains at this time, and
of this we give an illus-
tration (Fig. 47).

FIG. 47.—FROM MONUMENT OF
LYSICRATES, ATHENS.

The Corinthian order of architecture does not belong to
the early period of art in Greece. It came after the influence of Oriental architecture had been shown in the Ionic
style ; and perhaps the beautiful Corinthian capital may
have been suggested by the palm-leaf and lotus capitals of
Egypt. What has been said of other orders will help you
in understanding this ; but I shall tell you especially about
its capital, as that is its distinguishing feature. The form

of the capital may be called bell-shaped, and it is set round with two rows of leaves, eight in each row ; above these is a third row of leaves, or of a sort of small twisted husks, which supports eight small volutes. The abacus or top portion of the capital is cut out at the corners so that sharp projections are made, called horns, and one volute comes directly under each horn of the abacus. This cut (Fig. 48)

gives a more distinct idea of the capital than does that above, and you will see that four of the volutes really form the upper corners of the capital. The four other volutes meet on two opposite sides of the capital ; sometimes they are interwoven, and a flower, or rosette, or some other ornament is placed above them and lays up over the abacus. Different kinds of leaves are used in making this capital ; olive, water plant, and acanthus are all thus employed ; there is a very pretty legend as to its origin which makes the acanthus seem to be the only one which belongs to it, and is as follows :

FIG. 48.—CORINTHIAN ORDER.

It was the custom in Greece to place a basket upon the new-made graves in which were the viands which those there buried had preferred when in life. About 550 B.C. a lovely virgin died at Corinth, and her nurse arranged the basket with care and covered it with a tile. It happened that the basket was set directly over a young acanthus plant, and the leaves grew up about it in such a manner that the sculptor Callimachus was attracted by its grace and beauty, and conceived the idea of using it as a model for a new capital in architecture. I have always been sorry that it was not named for the beautiful maiden rather than for the city in which she was buried.

Another feature of Greek architecture is the use of the Caryatid, or a human figure standing upon a base and supporting the capital of a column upon the head, or, to put it more plainly, a human figure serving as the shaft to a column. These figures are usually females, and this picture of one from the Erechtheium at Athens shows how they are placed (Fig. 49). Sometimes the figures of giants, called *Telamones*, were used in the same way.

FIG. 49.—CARYATID.

In Oriental art such figures are numerous; they are used to support platforms and the

FIG. 50.—STOOL, OR CHAIR, KHORSABAD.

thrones of kings; their position is sometimes varied by making the uplifted hands bear the weight instead of the head (Fig. 50). In any case this feature in architecture is tiresome, and its use is certainly questionable as a matter of good taste.

Having given a general outline of the characteristics of Greek architecture, 1 will speak of some remarkable edifices which are beautiful in themselves and have an interest for us on account of their associations with the history of·the world, as well as with that of art.

The Temple of Diana at Ephesus, of which nothing now remains, was the largest and most splendid of all the Greek temples. It was four hundred and twenty-five feet long by two hundred and twenty wide.

The ancients counted this temple as one of the Seven Wonders of the World, and when we know that its pillars were sixty feet high, and that the beams of the architrave which had to be lifted up above the pillars to be put in place were each thirty feet long, we can readily understand that the building of it was a wonderful work. This was not the first temple that had stood on the same spot, for we know that one had been burned on the night in which Alexander the Great was born, 356 B.C. It was set on fire by Herostratus ; he was tried for this crime and was put to the torture to make him declare his motive for doing such a dreadful deed ; he gave as his only reason his desire to have his name handed down through all ages, and he believed that by burning the temple he should accomplish his object—as, indeed, he did, for every historian repeats the story of his crime, and his name stands as a synonym for wicked ambition.

After this destruction the temple was rebuilt on a most magnificent scale, and was not finished until two hundred and twenty years had passed. Diana was a great and powerful goddess, and all the nations of Asia united in gifts for the adornment of her shrine ; the women even gave their personal ornaments to be sold to increase the fund to be spent upon it.

This temple was four times as large as the Parthenon at Athens, and had one hundred and twenty-seven splendid columns, thirty-six of which were finely carved and were the gifts of various sovereigns. The grand staircase was made from the wood of a single Cyprian vine. But great as was the temple itself, its adornments of statues by the sculptor Praxiteles, and the vast treasures of ornaments and rare objects by which it was enriched made it even more famous. The Temple of Diana was robbed by Nero and burned by the Goths, but its final destruction probably occurred after A.D. 381, when the Emperor Theodosius I. issued an edict forbidding all the ceremonies of the pagan worship.

Many beautiful objects were taken away to adorn the mediæval churches of other religions than that of the Ephesians. Some of its green jasper columns were used to support the dome of St. Sophia at Constantinople, and other parts of it are seen in the cathedrals of Italy.

There is scarcely a more desolate spot in the world than is the Ephesus of to-day. No remaining ruins are so preserved as to afford the visitor any satisfaction. The marbles and stone have been used to build other towns, which in their turn have been destroyed. The inhabitants are a handful of poor Greek peasants ; wolves and jackals from the neighboring mountains roam about ; and though an abundance of myrtle and some lovely groves relieve the gloominess of the scene, it is impossible when there to re-create in imagination the splendid Ephesian city, with its wharves and docks, its temples, theatres, and palaces, which were so famous as to cause it to be spoken of with wonder throughout the ancient world.

We often hear of the glory of the Periclean age at Athens, and it is true that under the leadership of Pericles Athens reached its greatest prosperity. This picture shows the Acropolis as it appeared at that time (Fig. 51).

In these best days of Athens the whole Acropolis was consecrated to religious worship and ceremonials, and its entire extent was occupied by temples and statues of the gods. The fact that I have before mentioned, that the religion of a country moulds its art, is especially true of the art of Greece ; figures of the gods and bas-reliefs of the ceremonies of the Grecian worship form a large and most important part of the work of the Greek artists, and the splendid temples were raised to be the sacred homes of the statues of the great gods, to which the people could come with offerings and prayers.

The Acropolis was also a sort of fortress, because it was an eminence, and its sides of craggy rock allowed of but one ascent ; thus it could be easily defended. Then, when all the wonders and riches of art had been collected there, the pure white marble, the sculpture and painting, and the ornaments of shining metals which glistened in the sun, while brilliant colors added their rich effect, it might be called a gorgeous museum, such as has never since been equalled in the history of the world.

It is important to know that the Athenians worshipped three different goddesses, all called by the one name of Athene or Athena. The most ancient and most sacred of these was Athena Polias, whose statue, made of olive-wood, was believed to have fallen from heaven. The Erechtheium was dedicated to this goddess, and there this holy, heaven-sent figure was kept, with other sacred objects of which I shall speak in their place.

The Athena next in importance was the goddess of the Parthenon, or the " House of the Virgin," as the word signifies, for this Athena Parthenos is the same as the goddess Minerva, who is said never to have married or known the sentiment of love ; she was the goddess of war, prudence, and wisdom. The third Athena was called Promachos, which means the champion. Phidias made of her one of his splendid statues, standing erect, with helmet, spear, and shield.

In describing the Acropolis we shall begin with the Propylæa, or the entrances, which occupy the centre of our picture and to which the steps lead, showing the passage between the pillars, three being left on each side. This magnificent series of entrances—as the whole ascent

FIG. 51.—THE ACROPOLIS, *Athens.* (RESTORED.)

from the outer gate in the wall, up the steps, and through the passage between the pillars may be called—was erected about 437 B.C., and cost two thousand talents of gold, which is equal to about two millions of our dollars. The fame of the Propylæa was world-wide, and together with the Parthenon it was considered the architectural glory of the Periclean age. The style in which they are built is a splendid example of the combination of the Doric and the Ionic orders, for while the exterior is almost pure Doric, the interior is made more cheerful by the use of the Ionic columns and ornamentation.

High up at the right of the picture stands the Parthenon. Its architecture, which is Doric, has been described. We do not know when this temple was begun, but it is probably on the site of an older one. It was finished 438 B.C., and the general care of its erection was given to Phidias, the most famous of all sculptors. The marble of which the Parthenon was built was pure Pentelic, and as it rested on a rude basement of limestone the contrast between the two made the marble of the temple seem all the finer. Within and without this temple abounded in magnificent sculptures executed by Phidias himself or under his orders.

The Erechtheium, which is only partly visible at the back on the left of the picture, was the most sacred temple of Athens. It was the burial-place of Erechtheus, who was regarded not only as the founder of this temple, but also of the religion of Athena in Athens. Beside the heaven-descended statue of Athena Polias which was kept here, there was the sacred olive-tree which Athena had called forth from the earth when she was contending for the possession of Attica ; here, too, was the well of salt water which Poseidon (or Neptune) made by striking the spot with his trident, and several other sacred objects (Fig. 52).

This beautiful temple was built in the Ionic style, and is very interesting because it is so different in form from every other Greek temple of which we know. This is partly due to the fact that it was built where the ground was not level, one portion of it being eight feet higher than another. A second reason for its irregularity may be that it required to be divided into more cells or apartments than other Greek temples in order to arrange the different sacred objects within its walls. A very considerable portion of this temple is still standing. The frieze, of which but little remains, was of black marble, upon which there were figures in white marble.

The Erechtheium is certainly a splendid example of the Attic-Ionic style, and the eye rests upon it with admiration ; but its half-pillars and caryatides, its various porches and luxuriant detail of form and ornament, are less effective as a whole than is the Parthenon in its pure Doric architecture.

An interesting fact about Greek architecture is that the marbles used were painted in high colors. There is a theory, which may or may not be true, that the custom first arose in the same way as the shape of the Doric entablature, from the imitation of wooden buildings. The wood was painted to preserve it, and when stone began to be substituted, the architects, accustomed to bright effects, colored the marbles to look like wood. Whether this is the true origin of the custom or not, it is certain that the custom prevailed. The lower parts of the pillars of a Doric temple were usually stained a light golden-brown tint ; the triglyphs and the mutules, or brackets beneath the cornices, were a rich blue ; the trunnels, or wooden pins, were red or gilded ; the metopes had a dark red background, against which the bas-reliefs with which they were ornamented stood out in strong contrast, while the frieze and cornice were richly painted with garlands and leaves. So highly colored a building would seem less out of place amid the varied landscape of Greece than under our colder skies, and

FIG. 52.—THE ERECHTHEIUM. *Athens.* (RESTORED.)

it is difficult for us to form any just idea of the splendid appearance it must have presented.

One of the most wonderful things about Greek architecture is the way in which allowance was made for the deception of the eye by certain forms and lines. It is not easy to explain this fully, but it is too remarkable to be wholly passed over. If a column were cut so as to diminish regularly from the bottom to the top it would seem to the eye to hollow in, and to correct this the clever Greek architect made his columns swell out a little at the middle. This is called *entasis*, and is the best known of the means taken to make forms look as they should. Another case is that of long horizontal lines. If they are really level they appear to sag at the centre, therefore in Greek temples they are delicately rounded up a little, and so have the effect of being perfectly straight. These two examples may serve to show what I mean by saying that architectural forms were made one way so as to look another, and in nothing did the Greek architecture show more marvellous skill and taste than in this.

In other Grecian cities the architecture differed but little from that of Athens, and, indeed, the influence of Athenian art and artists was felt all over the Eastern world ; it is therefore not necessary for our purpose to speak further of Greek temples.

Next in importance were the municipal buildings, of which we find but few traces at Athens. The monument of Lysicrates is so beautiful that it gives us a most exalted idea of what the taste in such edifices must have been (Fig. 53).

This monument was erected in the year 334 B.C. when Lysicrates was *choragus;* this officer provided the chorus for the plays represented at Athens for the year. It was expensive to hold this position, and its duties were arduous ; the choragus had to find the men for the chorus, bring them together, and have them instructed in the music, and also provide proper food for them while they studied. It was customary to present a tripod to the *choragus* who provided the finest musical entertainment, and also to build a monument upon which the tripod was placed as a lasting honor to him who had received it. There was a street at Athens called the "Street of the Tripods" because it passed a line of choragic monuments. These monuments were dedicated to different gods; this of Lysicrates was devoted to Bacchus, and was decorated with sculptures representing scenes in the story of that god, who was regarded as the patron of plays and theatres; indeed, the Greek drama originated in the choruses which were sung at his festivals.

FIG. 53.—CHORAGIC MONUMENT OF LYSICRATES. *Athens.*

The Greek theatres were very large and fine; the seats were ranged in a half circle, but as none remain in a sufficient state of preservation to afford a satisfactory picture, it would be impossible to give a clear description of them here.

The ancient Greeks were not tomb-builders, and we know little of their burial-places. However, the Mausoleum built at Halicarnassus by Artemisia, in memory of her husband, Mausolus, was so important as to be numbered among the seven wonders of the world (Fig. 54).

FIG. 54.—THE MAUSOLEUM AT HALICARNASSUS (RESTORED).

Mausolus was the King of Caria, of which country Halicarnassus was the chief city. He died about 353 B.C., and
his wife, Artemisia, gradually faded away with sorrow at
his death, and survived him but two years. But during
this time she had commenced the erection of the Mausoleum, and the artists to whom she intrusted the work were
as faithful in completing it as though she had lived, for the
sake of their own fame as artists. This magnificent tomb
may be described as an example of architecture as a fine art

exclusively, for it cannot be said to have been useful, since the body of Mausolus was burned according to custom, and certainly a much smaller tomb would have been sufficient for the remaining ashes.

The whole height of the Mausoleum was one hundred and forty feet; the north and south aisles were sixty-three feet long, and the others a little less. The burial vault was at the base, and the whole mass above it was ornamented with magnificent designs splendidly executed. Above the whole was a quadriga, or four-horse chariot, in which it is said that a figure of Mausolus was placed so that from land or sea it could be seen at a great distance. It is not strange that this tomb was called a wonder in its day, and from it we still take our word "mausoleum" for all burial-places which merit so distinguished a name.

Writers of the twelfth century speak of the beauty of this tomb, but in A.D.1402, when the Knights of St. John took possession of Halicarnassus, it no longer remained, and a castle was built upon its site. The tomb had been buried, probably by an earthquake, and the name of the place was then changed to Boodroom.

In the year 1522 some sculptures were found there, but it was not until 1856 that Mr. Newton, an Englishman, discovered that these remains had belonged to the Mausoleum. A large collection of reliefs, statues, and other objects, more or less imperfect, was taken to London and placed in the British Museum, where they are known as the "Halicarnassus Sculptures."

As other temples were influenced by the example of the Athenian builders, so many other tombs resembled that of Mausolus in greater or less degree, although none approached it in grandeur and magnificence.

Of the domestic architecture of the Greeks we know very little. Almost all that is said of it is chiefly speculation, as even the descriptions of Grecian palaces and houses which are given by the classic writers are imperfect. The life of the Greek was passed largely in public, at the temple, the theatre, or the baths, or at least in the open air, and comparatively little attention was given to the building of the private houses ; but in the ruins of the temples and other monuments which still exist we have sufficient proof that no art has surpassed that of ancient Greece in purity, elegance, and grandeur of style.

ETRURIA.

Since the Etruscans were an earlier Italian nation than the Romans, and Rome, in her primal days, was ruled by Etruscan kings, it is here fitting to speak of this remarkable old people.

FIG. 55.—TOMBS AT CASTEL D'ASSO.

As Rome increased the Etruscans disappeared, and the younger power came to have so mighty an influence in the world that it absorbed the consideration of all nations as much as if no other had ever ruled in Italy.

No Etruscan temple now remains, but we know that they were not splendid like those of Greece. They were of two forms, one being circular and dedicated to a single deity, while others were devoted to three gods and had

FIG. 56.—PRINCIPAL CHAMBER IN REGULINI-GALEASSI TOMB.

three cells; their walls were built at right angles, thus making their shape regular.

The theatres and amphitheatres of the Etruscans were nearly circular and much like those of the later Italians, but not one remains except that at Sutri, which, being cut in the rock, does not afford a good example of the usual arrangement of these edifices.

In fact, the only important remains of Etruscan architecture are the tombs, of which there are many. These are of two kinds; the first are cut in the rocks and resemble the Egyptian tombs at Beni-Hassan, reminding one of little houses (Fig. 55).

The second and most numerous class are mounds of earth raised above a wall at the base. These were called "Tumuli," and some of them had fine, well-furnished apartments in their midst. The next cut shows such a room as it appeared

FIG. 57.—ARCH AT VOLTERRA.

when first opened; in it were found bedsteads, biers, shields, arrows, a variety of vessels, and several kinds of useful utensils (Fig. 56).

FIG. 58.—GATEWAY. *Arpino*.

These tombs are in truth more connected with other arts than with architecture, and many beautiful articles have been found in them. The most interesting feature of Etruscan architecture is the arch, which was first brought into general use by the Romans, but is found in Etruscan remains (Fig. 57), both in the semi-circular and pointed forms. The principle of the arch had been known to several Oriental nations, but it had been applied only to short spaces and comparatively unimportant uses, such as windows and doorways (Fig. 58).

There is no doubt that many of the earliest works of the Romans were executed under the direction of Etruscan architects. Among these was the great Cloaca Maxima, or

FIG. 59.—ARCH OF CLOACA MAXIMA. *Rome.*

principal drain of ancient Rome. This was a wonderful achievement ; it is probable that the oldest arch in Europe is that of this sewer, and the fact of its still remaining proves how well it must have been built in order to last so long (Fig. 59).

ROME.

The early works of Rome, which were largely executed by the Etruscans, were principally those useful, semi-archi-tectural objects necessary in the making of a city, such as aqueducts and bridges.' These belong quite as much to civil engineering as to architecture, and we shall not speak of them.

In studying Roman architecture one is surprised at the number of uses to which it was applied, for not only do

the temples, tombs, theatres, and monuments such as we
have found in other countries exist in Rome, but there are
also basilicas, baths, palaces, triumphal arches, pillars of
victory, fountains, and
various other objects
suited to the wants of
a great people.

No truly pure,
national order of ar-
chitecture existed at
Rome. The union of
the arch of the Etrus-
cans with the columns
of the Greeks enabled
the Romans to change
the forms of their edi-
fices and to produce a
great variety in them.
They employed the
Doric, Ionic, and Co-
rinthian orders, but
they rarely used one
of these alone ; they
united them in endless
combinations, and in-
troduced a capital of
the order which is
called the Composite
(Fig. 60). It consists

FIG. 60.—COMPOSITE ORDER, FROM THE
ARCH OF SEPTIMIUS SEVERUS. *Rome.*

of the lower part of the Corinthian and the upper part of
the Ionic capital ; this was very rich in ornament, but the
line where the two orders were joined was always a defect,
and it never came into general favor.

The Romans also introduced what is called the Tuscan
order, which is usually mentioned with the Doric, Ionic,

Corinthian, and Composite, as being one of the five classic
orders of architecture, although it is really little more than
a variety of the Doric, as the Composite is of the Corinthian
order.   It differed from the Doric in having a base, while
its frieze was simple and unadorned, the cornice also being
very plain.   The shaft of the Tuscan column was never
fluted.

The Romans also used an arcade which was a combination of Greek and Etruscan art, like this cut (Fig. 61); thus
showing a power of adapting forms which already existed
in new combinations and for
new purposes, rather than an
originative genius.

A very important advance
made by the Romans was the
improvement of interior architecture. The halls and portions of edifices to be used
were more cared for than ever
before ; this was sometimes
done at the expense of the
exteriors, to which the Greeks
had devoted all their thought.
In fact, many ancient Roman
temples were inferior to other

FIG. 61.—DORIC ARCADE.

edifices which they built.   The Pantheon is the only one
existing in such a state as to be spoken of with satisfaction.

This ground-plan (Fig. 62) shows that the Pantheon is
circular with a porch.   Taken separately, the rotunda and
the porch are each fine in their own way, but the joining
of the circular and angular forms has an effect of unfitness
which one cannot forget even when looking at that which
we regard with reverent interest.   The central portion was
at first a part of the Baths of Agrippa, but on account of
its great beauty it was changed by Agrippa himself into

a temple, by the addition of a row of Corinthian columns
around the interior.   (See Fig. 63.)

Taken all in all, the effect of the Pantheon is that of
grandeur and simplicity.   When we remember that sixteen
hundred and eighty-eight years have passed since it was

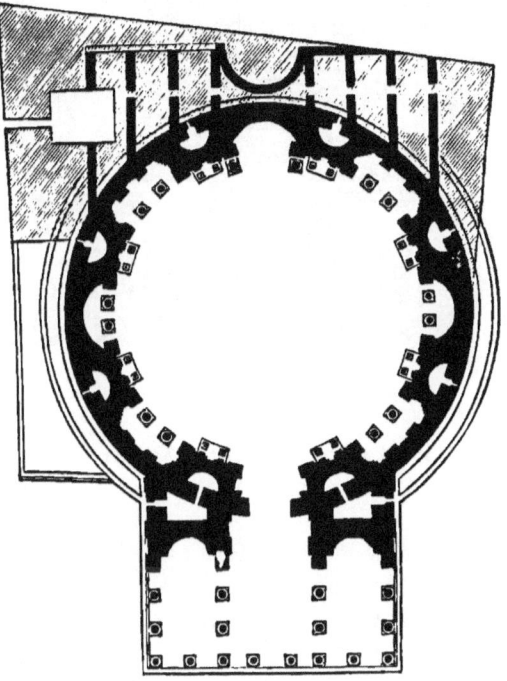

FIG. 62.—GROUND-PLAN OF PANTHEON.   *Rome.*

repaired by Septimius Severus, we wonder at its good preservation, though we know that it has been robbed of its
bronze covering and other fine ornaments.   An inscription
still remaining on its portico states that Marcus Aurelius
and Septimius Severus repaired this temple ; history says
that Hadrian restored it after a fire, probably about the

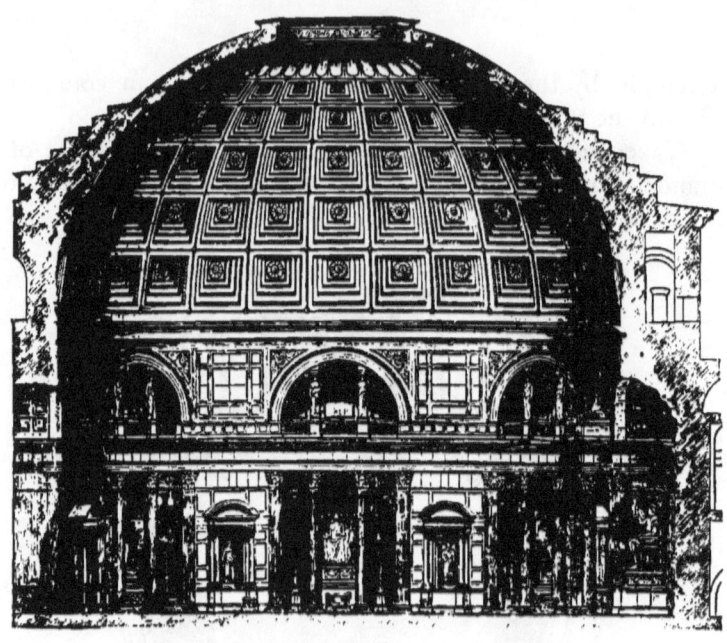

FIG. 63.—INTERIOR OF THE PANTHEON.

year 117, and it is even said that Agrippa, who died A.D. 13, added the portico to a rotunda which existed before his time.

The objects now in the interior of the Pantheon are so largely modern that they do not belong to this portion of our subject, but there is much interest associated with this spot, and it is dear to all the world as the burial-place of Raphael, Annibale Caracci, and other great artists.

Next to the temples of Rome came the Basilicas, of which there were many before the time of Constantine. The word basilica means the royal house, and these edifices were first intended for a court-room in which the king administered his laws ; later they became markets, or places of exchange, where men met for business transactions.

The ruins of the Basilicas of Trajan and Maxentius, two of the finest of these edifices, are in such condition that their plans can be understood (Fig. 64). They were large, and divided into aisles by rows of columns ; at one end there was a semi-circular recess or apse, in which was a raised platform, approached by steps, also semi-circular in form. Upon this platform the king or other exalted officer had his place, while those of lesser rank were on the steps below, on either side. Fronting the apse was an altar upon which sacrifices were offered before commencing any important business.

FIG. 64.—LONGITUDINAL SECTION OF BASILICA OF MAXENTIUS.

The principal reason for speaking of basilicas is that by the above cut you may see the great change made in archi-tecture about this time by the use of columns, only half the height of the building, which were united by arches. This was a very important step, and is, in truth, one of the principal features that mark the progress of the change from ancient to Gothic architecture—a change not fully developed until the twelfth century.

I shall not say much of the theatres, amphitheatres, and baths of ancient Rome, because it is not easy to treat them in the simple manner suited to this book ; they were mag-nificent and costly, and made an important part of Roman architecture ; they were probably copied from the public buildings of the Etruscans.

Marcus Scaurus built a theatre in 58 B.C. which held eighty thousand spectators ; it had rich columns and statues, and was decorated with gold, silver, and ivory. The first stone theatre in Rome was built in 55 B.C., and was only half the size of that of Marcus Scaurus. Parts of the theatre of Marcellus still remain in the present Orsini Palace in Rome, and serve to give an idea of the architecture of the period immediately before the birth of Christ.

The Emperor Augustus boasted that he had found a city of brick and had changed it to one of marble, but after his time architecture suffered a decline, and its second flourishing period may be dated from A.D. 69. To this time belongs the Colosseum, also called the Flavian Amphitheatre ; it covers about five acres of ground, and is sufficiently well preserved for a good idea to be formed of what it must have been when in its best estate. The enormous size of these ancient Roman edifices is almost too much for us to imagine, and the most extensive of them all were the *Thermæ*, or public baths.

The Baths of Diocletian, built A.D. 303, were the largest of all ; they had seats for twenty-four hundred bathers. These baths were in reality a group of spacious halls of varied forms, but all magnificent in size. The great hall of the Baths of Diocletian was three hundred and fifty feet long by eighty feet in width and ninety-six feet high ; it was converted into a church by Michael Angelo and is called S. Maria Degli Angeli, or Holy Mary of the Angels. Many splendid pictures which were once in St. Peter's are now in this church, and copies of them made in mosaic fill the places where they were originally hung.

The Baths of Caracalla were built in A.D. 217, and though they had seats for but sixteen hundred bathers, they were much more splendid than the Baths of Diocletian.

They were surrounded by pleasure gardens, porticoes, and a stadium or race-course, where all sorts of games were held. Some beautiful mosaic pavements have been taken from these baths, and are now in the Lateran and the Villa Borghese palaces; there was a Pinacotica, or Fine Art Gallery here, in which were some of the greatest art treasures of the world, such as the Farnese Hercules, the Farnese Bull, the two Gladiators, and other famous statues, besides cameos, bronzes, and sculptures, almost without end. The granite basins in the Piazza Farnese, and some green basalt urns now in the Vatican Museum, were taken from the Baths of Caracalla, and, indeed, all over Rome there are objects of more or less beauty which were found here.

Formerly the site of these baths was like a beautiful Eden where Nature made herself happy in luxuriant growths of all lovely things. The poet Shelley was very fond of going there, and wrote of it, "Among the flowery glades and thickets of odoriferous blossoming trees, which are extended in ever-winding labyrinths upon its immense platforms and dizzy arches suspended in the air," by which we know that the ruins were covered with a soil which was fruitful in flowers, vines, and trees; but all these have been torn away in order to make the excavations which were necessary for the exploration of these wonderful baths, and now the parts which remain stand fully exposed to the view of the curious traveller.

The Roman Triumphal Arches were one of the characteristic outgrowths of the Imperial period. These splendid works were designed to perpetuate the fame of the emperors and to recall to the people the important acts of their lives. The arch of Constantine given below is one of the most famous arches in Rome (Fig. 65). It is believed that parts of it were in an arch of Trajan's time, and some even go so far as to say that it was originally dedicated ·to

the earlier emperor and adopted by Constantine as his own.
It is remarkably well preserved, and this is undoubtedly due
to the fact of its being dedicated to the first Christian sovereign of Rome.   The other most famous arches in the city
are that of Titus, which dates from A.D. 81, and that of
Septimius Severus, which was erected in honor of him and

FIG. 65.—ARCH OF CONSTANTINE.   *Rome.*

of his wife, Julia, by the silversmiths and merchants of the
Forum Boarium, in which spot the arch was raised.

These triumphal arches existed in all the countries
where Rome held sway, and, indeed, this is true of all
kinds of Roman architectural works.

This Arch of Beneventum was erected in the second
century after Christ, by Trajan, when he repaired the

Appian Way. It is one of the most graceful and best preserved of all the arches of Italy (Fig. 66).

All these arches had originally groups of statuary upon
them, for which they served merely as the pedestals. Their
taking the form of an arch was due to their being placed in
the public way, where it was necessary to leave a passage

FIG. 66.—ARCH OF TRAJAN. *Beneventum.*

for the street. Sometimes they were placed where two
roads met, and a double arch was then made. Elaborate
as the arches often were, you must keep in mind that they
are only a part of the entire design, and that the least
important part ; the statuary, which has been destroyed by
time, being really the more striking feature of the whole.

The tombs of Rome were very numerous, and were an

important element in Roman architecture. The tomb of Cecilia Metella is of importance because it is the oldest

remaining building of Imperial Rome and the finest tomb which has been preserved (Fig. 67).

As you see, the tomb is a round tower. In the thirteenth century it was turned into a fortress, and so much dust has been deposited on its summit in the passing of time that bushes and ivy now grow there. Many writers describe it, and Byron in his "Childe Harold" spoke of it in some verses, of

FIG. 67.—TOMB OF CECILIA METELLA.

which the following is the beginning :

"There is a stern round tower of other days,
Firm as a fortress, with its fence of stone,
Such as an army's baffled strength delays,
Standing with half its battlements alone,
And with two thousand years of ivy grown,
The garland of eternity, where wave
The green leaves over all by time o'erthrown ;—
What was this tower of strength ? within its cave
What treasure lay so lock'd, so hid ?—a woman's grave."

The tomb of Hadrian, now known as the Castle of St. Angelo, is very interesting, and is one of the most prominent and familiar objects in Rome at the present day. But the tombs called Columbaria were much in use in ancient Rome, and differed essentially from those of which we have spoken, inasmuch as they were usually below the ground, and externally had no architecture. They consisted of oblong or square apartments, the sides of which were filled with small apertures of the proper size to hold an urn which contained the ashes that remained after a body had been burned, according to the Roman custom. Some of these apartments, especially when they belonged to private families, were adorned with pilasters and decorated with colors. (See Fig. 68.)

FIG. 68.—COLUMBARIUM NEAR THE GATE OF ST. SEBASTIAN. *Rome.*

The sepulchres of Rome were gradually enlarged, until, in the days of Constantine, they were frequently built like small temples above the ground, with crypts or vaults beneath them.

So little now remains of the ancient domestic architecture of Rome that one is forced to study this subject from written descriptions collected from the works of various historians, poets, and other writers. But from what we know we may conclude that the villas and country-houses were so constructed as to be full of comfort, and suited to the uses for which they were built, without too much regard to the symmetry of the exteriors. The interior convenience was the chief thing to be considered, and when finished they must have often resembled a collection of buildings all joined together, of various heights and shapes ; but within they were adapted to the different seasons, as some rooms were made for being warm, while others were arranged for coolness ; the views from the windows were also an important feature, and, in short, the pleasure of the people living in them was made the first point to be gained, rather than the impression upon the eye of those who saw them from without.

There was great luxury and elegance in the palaces of the noble classes in ancient Rome. The home of Diocletian at Spalatro was one of the most famous Roman palaces, and its ruins show that it was once magnificent. This palace was divided by four streets which ran through it at right angles with each other and met in its centre. Its entrances were called the Golden, Iron, and Brazen Gates. Its exterior architecture was simple and massive, as it was necessary that it should serve as a fortress in case of an attack. Its principal gallery overlooked the sea ; it was five hundred and fifteen feet long and twenty-four feet wide, and was famous for its architectural beauty and for the views which it commanded.

CHAPTER II.

CHRISTIAN ARCHITECTURE.

A.D. 328 TO ABOUT 1400.

I HAVE written more in detail concerning Ancient architecture than I shall do of that of later times, because it is best to be thorough in studying the beginnings of things ; then we can make an application of our knowledge which helps us to understand the results of what has gone before, just as we are prepared for the full-blown rose after we have seen the bud. Or, to be more practical, just as we use the simplest principles of arithmetic to help us to understand the more difficult ones ; sometimes we scarcely remember that in the last lessons of the book we unconsciously apply the first tables and rules which were so difficult to us in the beginning.

I shall not try, because I have not space, to give a connected account of Christian architecture, but I shall endeavor to give such an outline of its rise and progress in various countries as will make a good foundation for the knowledge you will gain from books which you will read in future.

The architecture of Italy in the period which followed the conversion of the Emperor Constantine is called the Romanesque order. As the Christians were encouraged under Constantine and became bold in their worship, many basilicas were given up for their use. The bishops held the

FIG. 69.—INTERIOR OF BASILICA OF ST. PAUL'S. *Rome.*

principal place upon the platform formerly occupied by the
king and his highest officers, and the priests of the lower
orders were ranged around them.   The same altars which
had served for the heathen sacrifices were used for the worship of the true God, and from this cause the word basilica
has come to signify a large, grand church, in the speech of
our time.

Among the early basilicas of Rome which still remain
none are more distinguished than that of *San Paolo fuori
della Mura*, or St. Paul's without the Walls.   It was ancient,
and splendid in design and ornament.   In 1823 it was
burned, and has been rebuilt with great magnificence, but
the picture above shows it as it was before the fire (Fig. 69).
It was built about 386 A.D. under the Emperors Valen-
tinian II. and Theodosius.

This basilica had four rows of Corinthian columns, twenty in each row ; many of these pillars were taken from more ancient edifices, and were composed of very beautiful marbles, forming by far the finest collection of columns in the world. The bronze gates were cast at Constantinople ; the fine paintings and magnificent mosaics with which it was decorated added much to its splendor. Tradition taught that the body of St. Paul was buried beneath the high altar.

Before the Reformation the sovereigns of England were protectors of this basilica just as those of France were of St. John Lateran ; this gives it a peculiar interest for British people, and the symbol of the Order of the Garter is still seen among its decorations. On account of its associations, San Paolo was the most interesting, if not the most beautiful, of the oldest Christian edifices in Rome.

In the early days there were many circular churches throughout Italy ; some of these had been built at first for tombs. The Christians used churches of this form for baptisms, for the sacrament for the dying, burials, and sometimes for marriage.

The circular temple of Vesta is very beautiful. It had originally twenty Corinthian columns ; nineteen of which still remain. This temple is not older than the time of Vespasian, and is not the famous one mentioned by Horace and other ancient writers, in which the Palladium was preserved—that temple no longer exists. It is probable that many of the earliest churches built by Christians in Italy were circular in form, and numbers of these still remain in various Italian cities ; but they differed from the ancient temples of this form in their want of exterior decoration. The ancient Romans had used columns, peristyles, and porticoes ; the Christians used the latter only in a few instances, but even these were soon abandoned.

The beautiful Baptistery at Florence was originally the

cathedral of the city. It is octagonal, or eight-sided, and this form is not infrequent in buildings of the fourth and following centuries. It is said that this Baptistery was built by Theodolinda, who married Autharis, King of the Lombards in 589.

This king had proposed to Garibald, King of Bavaria, for the hand of his daughter, and had been accepted. Autharis grew impatient at the ceremonies of the wooing, and escaping from his palace joined the embassy to the King of Bavaria.

When they reached the court of Garibald and were received by that monarch, Autharis advanced to the throne and told the old king that the ambassador before him was indeed the Minister of State at the Lombard Court, but that he was the only real friend of Autharis, and to him had been given a charge to report to the Italian king concerning the charms of Theodolinda. Garibald summoned his daughter, and after an admiring gaze the stranger hailed her Queen of Italy and respectfully asked that she should, according to custom, give a glass of wine to the first of her future subjects who had tendered her his duty. Her father commanded her to give the cup, and as Autharis returned it to her he secretly touched her hand and then put his finger on his own lips. At evening Theodolinda told this incident to her nurse, who assured her that this handsome and bold stranger could have been none other than her future husband, since no subject would venture on such conduct.

The ambassadors were dismissed, and some Bavarians accompanied the Lombards to the Italian frontier. Before they separated Autharis raised himself in his stirrups and threw his battle-axe against a tree with great skill, exclaiming, " Such are the strokes of the King of the Lombards !" Then all knew the rank of this gallant stranger. The approach of a French army compelled Garibald to leave his

FIG. 70.—THE CATHEDRAL OF CHARTRES.

capital ; he took refuge in Italy, and Autharis celebrated his marriage in the palace of Verona ; he lived but one year, but in that time Theodolinda had so endeared herself to the people that she was allowed to bestow the Italian sceptre with her hand. She had converted her husband to the Catholic faith. She also founded the cathedral of Monza and other churches in Lombardy and Tuscany, all of which she dedicated to St. John the Baptist, who was her patron saint.

The cathedral of Monza is very interesting from its historical associations. Here is deposited the famous iron crown which was presented to Theodolinda by Pope Gregory I. This crown is made of a broad band of gold set with jewels, and the iron from which it is named is a narrow circlet inside, said to have been made from one of the nails used in the crucifixion of Christ, and brought from Jerusalem by the Empress Helena. This crown is kept in a casket which forms the centre of the cross above the high altar in the cathedral of Monza ; it was carried away in 1859 by the Austrians ; at the close of the Italo-Prussian war, in 1866, the Emperor of Austria gave it to Victor Emmanuel, then King of Italy. This crown has been used at the coronation of thirty-four sovereigns ; among them were Charlemagne, Charles V., and Napoleon I. The latter wore it at his second coronation as King of the Lombards in 1805. He placed it on his head himself, saying, " God has given it to me, woe to him who touches it !"

There are few secular buildings of this period remaining in Italy, and Romanesque architecture endured but a short time, for it was almost abandoned at the time of the death of Gregory the Great, in 604. During the next four and a half centuries the old styles were dying out and the Gothic order was developing, but cannot be said to have reached any high degree of perfection before the close of the eleventh century.

GOTHIC ARCHITECTURE.

It is difficult to speak concisely of Gothic architecture because there is so much that can be said of its origin, and then it has so extended itself to all parts of the world as to render it in a sense universal. Perhaps Fergusson makes it as simple as it can be made when he divides Europe by a line from Memel on the shores of the Baltic Sea to Spalatro on the Adriatic, and then carries the line westward to Fermo and divides Italy almost as the forty-third parallel of latitude divides it. He then says that during the Middle Ages, or from about the seventh to the fifteenth centuries, the architecture north and west of these lines was Gothic ; south and east it was Byzantine, with the exception of Rome, which always remained individual, and a rule unto herself.

There was a very general belief in all Christian lands that the world would end in the year 1000 A.D., and when this dreaded period had passed without that event happening, men seem everywhere to have been seized with a passion for erecting stone buildings, An old chronicler named Rodulphe Glaber, who died in 1045 A.D., relates that as early as the year 1003 A.D. so many churches and monasteries of marble were being erected, especially in France and Italy, "that the world appeared to be putting off its old dingy attire and putting on a new white robe. Then nearly all the bishops' seats, the churches, the monasteries, and even the oratories of the villages were changed for better ones."

Such a movement could not fail to have a great influence upon architecture, and it was at this time that the Gothic style began to be rapidly developed ; and, indeed, so far as any particular time may be fixed for the beginning of the Gothic order, it would fall in the tenth and eleventh centuries. The classic forms, with their horizontal cornices and severe regularity, were then laid aside, and a greater freedom and variety than had ever obtained before began to make itself felt in all architectural designs.

We must first try to understand what are the distinguishing features of Gothic architecture. Perhaps the principal one may be called constructiveness; which is to say, that in Gothic architecture there is far greater variety of form, and the power to make larger and more complicated buildings than had been possible with the orders which preceded it. During the Middle Ages the aim was to produce large edifices, and to build and ornament them in a way that would make them appear to be even larger than they were. The early Gothic buildings are so massive as to have a clumsy effect, because the architects had not yet learned how to make these enormous masses strong and enduring, and yet so arranged as to be light and graceful in their appearance.

A second striking difference between the ancient orders and the Gothic, is that in the former enormous blocks of stone or marble were used and great importance was attached to this. Many ancient works are called Cyclopean for this reason. It does not make a building more beautiful to have it massive, but it does make it grand. Even in a less colossal mode of building a column is more effective when it is a monolith, and an architrave more beautiful when its beams are not joined too frequently. But in the Gothic order the use of massive blocks is largely given up, and the endeavor is to so arrange smaller materials as to display remarkable constructive skill.

A third and a very important feature of the Gothic order is the use of the arch. The much-increased constructive power of which we have spoken depended very largely upon this. The ancients knew the use of the arch, but did not like it because they thought that it took away from the

FIG. 71.—CHURCH OF ST. NICHOLAS. *Caen.*

repose of a building. Even now the Hindoos will not use it ; they say, "An arch never sleeps," and though the Mohammedan builders have used it in their country, the Hindoos cannot overcome their dislike of it. In the Gothic order, however, the use of arches, both round and pointed, is unending. The results are very much varied, and range all the way from a grand and impressive effect to a sort of toy-like lightness which seems more suited to the block-houses made by children than to the works of architects. The earlier Gothic arches were round, although pointed

FIG. 72.—FAÇADE OF CATHEDRAL OF NOTRE DAME. *Paris.*

arches are occasionally found in very ancient buildings.
The picture (Fig. 71), however, gives a just idea of the
form of arch most used until the introduction of the
pointed arch, which occurred in France during the twelfth
century. Of this form the doorways of the next cut present a fine example (Fig. 72).

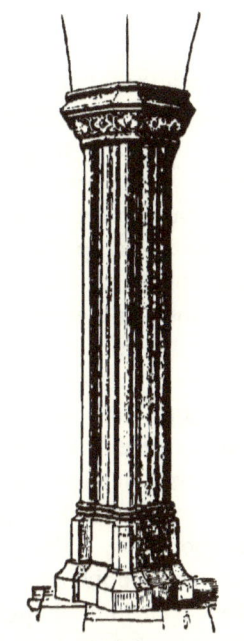

FIG. 73.—CLUSTERED PILLAR.

FIG. 75.—HINGE.

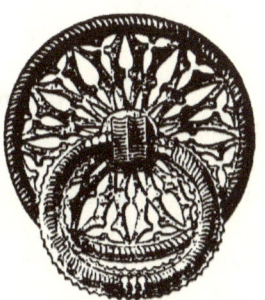

FIG. 77.—IRON-WORK.

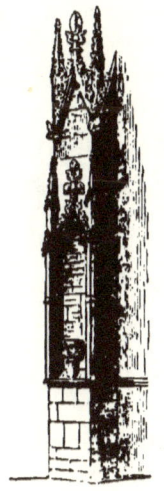

FIG. 74.—BUTTRESS.

FIG. 78.—GARGOYLE.

An important characteristic of Gothic architecture was the fact that every part of the building was so made as to show its use. Instead of hiding the supports they were made prominent. If a pier or buttress was to stand a perpendicular strain, even the lines of decoration were generally made to run in that direction ; if extra supports were

needed, they were not concealed, but built in so as to show, and even to be prominent. In the details the same feeling was often shown in a very marked degree ; the hinges and nails and locks of Gothic buildings were made to be seen, and whatever was needed for use was treated as if it were of value as an ornament. The spouts by which the

FIG. 76.—NAIL-HEAD.

water was carried over the eaves were made bold and comparatively large, and carved into those curious shapes of animals and monsters called gargoyles, which are seen on so many mediæval edifices. Many of these details of Gothic buildings are very elegant, and serve to-day as models for modern workmen. (See Figs. 73, 74, 75, 76, 77, 78, 79.)

Among the inventions of Gothic architects the division of the interior into three aisles, with the centre one much the highest, was very important. By this arrangement the space was made to appear longer and higher than it really was, and what was lost in the effect of

FIG. 79.—SCROLL.

width was more than made up in a certain elegance of form which is very pleasing. The three central aisles of the next cut illustrate this arrangement (Fig. 80).

The Gothic builders gave loftiness to their edifices by the use of spires and towers. They became very skilful in constructing them with buttresses below and pinnacles above, so that the spires should not detract from the apparent size of the buildings to which they were attached (Fig. 81).

In the matter of design in ornament the Gothic order had no fixed method, except so far as its forms were symbolic. Every form of vegetable design was employed ; vines and leaves were abundant. As a rule the use of human forms or animals as supports to columns or other weights

FIG. 80.—SECTION OF CHURCH. *Carcassone.* WITH OUTER AISLES ADDED IN FOURTEENTH CENTURY.

was avoided. If they were introduced the animals were not reproductions of such as exist, but the imaginary griffin or other monster, and at times dwarfs or grotesque human beings, were represented as if for caricatures.

Sculptured figures were usually placed upon a pedestal either with or without niches for them, and were not made to appear to be a part of the building itself. The deep recesses of Gothic portals, the pinnacles and niches gave

FIG. 81.—SPIRES OF LAON CATHEDRAL.

opportunities to display exterior sculpture to great advantage (Fig. 82). The interiors were also appropriate for any amount of artistic ornament in bas-reliefs or figures that could be lavished upon them.

The most original and effective feature of ornament, however, which was introduced by Gothic architects is that of painted glass. To this they devoted their best talent. It is not necessary to say how beautiful and decorative it is ; we all know this, and our only wonder is that it was left for the Gothic architects to apply it to architectural uses. We do not know precisely when stained or painted glass was invented, but we know that it existed as early as 800, and came into very general use in the eleventh and twelfth centuries.

Before painted glass was used windows were made very small, and it was some time before the large, rich style was adopted. The following cut from Notre Dame, at Paris, gives the three stages of the change, and it is interesting to see them thus in one church (Fig. 83).

On the left are the undivided windows without mullions or dividing supports ; next, at the right, the upper window shows the form with one perpendicular mullion and a circular or rose window above the centre ; lastly, on the right of the lower story we see a full traceried window.

The window became one of the most important and characteristic features of Gothic buildings. These large

FIG. 82.—PORTAL OF THE MINORITES' CHURCH. *Vienna.*

open spaces gave opportunity for elegant shapes and splendid colors, both the form of the opening and the dividing ribs, or tracery, as it was called, being made with the utmost beauty and grace. The round windows, called rose windows and wheel windows, were often exquisitely designed, as the following example shows (Fig. 84).

The window is illustrative of the influence which climate may have on the development of architectural style. In warm countries where spaces were left open, window forms and painted glass were, of course, never employed ; but in more northern lands they became one of the most marked features in important edifices.

A whole book might be written about these windows and be very interesting also, but we can give no more space to them here.

FIG. 83.—EXTERNAL ELEVATION, CATHEDRAL OF PARIS.

Gothic architecture gradually extended from the centre of Italy to the most northern bounds of civilization, and though practised by so many nations, was as much the architectural expression of a religion as the architecture of a single ancient nation had been the outgrowth of its peculiar religious belief. During the Middle Ages the priests and monks preserved learning in the midst of general darkness and ignorance, and were the chief patrons of all art which survived the decline of the time. They built up the Christian faith by every means in their power. The monks were missionaries. They went to various countries, and selecting favorable spots they founded abbeys ; around these abbeys a poor population settled ; gradually churches were built, and it frequently happened that the monks not only planned the work to be done, but also executed it with their own hands. Many of them were masons and builders, and several bishops were architects. St. Germain, Bishop of Paris, designed the church in that city now called by his name, and was also sent to Angers to build another church, and to Mans to erect a monastery.

FIG. 84.—WHEEL WINDOW, FROM CATHEDRAL. *Toscanella.*

The finest buildings being thus made for religious purposes and under the direction of the clergy, they must have been as full an expression of Christianity as were the temple-palaces of Egypt an expression of the religion of Osiris and Isis, when the kings were both priests and sovereigns, and dwelt in these palaces. And this was true as long as Gothic art was in the hands of the clergy and used almost entirely for religious purposes.

Later on, when it was employed for civic edifices erected under the direction of laymen, it became an expression of political independence also. The freedom of thought which came with the decline of the feudal system inspired new aspirations and imaginations in the hearts and minds of men, and these found expression in all the arts, and very especially in architecture. If we cannot always admire the manner in which Gothic art was made to express these lofty desires, we can fully sympathize with the sentiment which was behind it.

The Gothic order held undisputed sway west and north of the geographical line of which we have spoken until the fifteenth century. Then a revival of classical literature took place, and with this there arose also a revival of classic art and architecture ; this revival is known as the Renaissance, or the new birth, and the period of time is spoken of as that of the Renaissance. The effect of this classic reaction was very great upon all the educated classes of Europe, and its influence may be said to have endured through about three centuries.

Again, during the eighteenth century, Gothic art was revived. A reverence has grown up for the good that wrestled with the darkness of the Middle Ages and survived all their evils. The rough, strong manhood of that time is now justly appreciated. Perhaps the feeling in this direction is too much exaggerated. While our regard for a rude and weather-stained monument of the spirit and architecture of the past may be natural and proper, the imitation of it which is made in our day may easily become absurd, and is very rarely suited to our purposes.

Spain is one of the countries which are on the Gothic side of the geographical line we have drawn, and among the many splendid edifices in that country some of the finest are of the Gothic order. There is no national architecture there, for though the Spaniards love art and its expression

FIG. 85.—COLLEGIATE CHURCH, TORO. *From Villa Amil.*

passionately, they have themselves invented almost nothing which is artistic.

But while it is true that the Spaniards invented no styles, they did modify those which they adopted, and there are peculiarities in the Spanish use and arrangement of the Gothic order which give it new elements in the eyes of those who understand architecture scientifically. To the uneducated also it appears to have a personality of its own, something that is suited to Spain and the Spaniards ; so that, while we know that Spanish Gothic architecture was borrowed from France and Germany, we yet feel that if the cathedrals of Paris and Cologne were to be put down in Valencia or Madrid they would look like strangers, and not at all well-contented ones at that ; and if the churches of Toledo or Burgos were copied precisely in any other country, they would have an air of being quite out of keeping with everything around them (Fig. 85).

We call the architecture of Spain before 1066 the " Early Spanish," and from that time the Gothic order prevailed during nearly three centuries.

FIG. 86.—ST. PAUL. *Saragossa.*

Meantime in the south of Spain the Moresco or Moorish order had sprung up, of which Fig. 86 gives an example. It was gradually adopted to a limited extent, until finally some specimens of it existed in almost every province of the country. The Gothic order was affected by it, inasmuch as the richness of ornament of the Moorish order so pleased the taste of the Spaniards that their architects allowed themselves to indulge in a certain Moorish manner of treating the Gothic style. We cannot describe these differences in words, but Figs. 86 and 87 will make it plain.

As has been said, the interior decoration of all Gothic churches was very rich and abundant. It is also true that all church furniture was made with great care ; the matter of symbolism was carefully considered, and each design made to indicate the use of the article for which it was intended. No altar, preaching-desk, stall, chair, or screen was made without due attention to every detail, and the endeavor to have it in harmony with its use and its position in the church. The following cut shows a rood-screen, which was the kind of screen that was placed before the crucifixion over the high altar (Fig. 88).

The fantastic sculptures and wealth of ornament in Gothic decorations pro-duce a confusing effect on the brain and the eye if we look at the whole carelessly ; but when we remember that each separate design has its especial meaning we are interested to exam-ine them, and we find that the variety of forms is almost innumerable. Where there are trailing vines and lions, faith is indicated ; roses and pelicans are the symbols of mercy and divine love ; dogs and ivy, of truth ; lambs, of gentie-ness, innocence, and sub-mission ; fishes are an emblem of water and the rite of baptism ; the

FIG. 87.—CLOISTER. *Tarragona.*

dragon, of sin and paganism ; a serpent, too, typifies sin, and when wound around a globe it indicates the power of evil over the whole world ; a hind or hart signifies solitude ; the dove, purity ; the olive, peace ; the palm, martyrdom ; the lily, purity and chastity ; the lamp, lantern, or taper, piety ; fire and flames, zeal and the suffer-

FIG. 88.—ROOD-SCREEN, FROM THE MADELEINE. *Troyes.*

ings of martyrdom ; a flaming heart, fervent piety and
spiritual love ; a shell, pilgrimage ; a standard or banner,
victory ; and so on, and on, we find that meaning and
thought were worked out in every bit of Gothic ornament,
and that what at first appears so wild and hap-hazard is full
of a method which well repays one for the study of it.

The Gothic order was also used in building municipal

FIG. 89.—PALACE OF WARTBURG.

edifices, palaces, and even for the purposes of domestic
architecture. The finest remains of this kind are in Germany, the most interesting of them all being the castle on
the Wartburg. This castle is large, grand, and imposing.
It is also well preserved. A few years ago it was discovered
that many windows and arched galleries, of very beautiful
style, had been filled up, and that frescoes and other deco-

rations had been covered. The Grand Duke of Saxe-Weimar caused its restoration, and the ancient halls are now quite in their original state. (See Fig. 89.)

There are very interesting legends and historical facts connected with this castle of Wartburg. As early as 1204 to 1208, when Hermann, Count of Thuringia, dwelt there with his wife, the Countess Sophia, it is related that the " War of the Minstrels" occurred. This was a contest between several of the wandering minstrels or Minnesingers of that time as to who should excel, and he who failed was to suffer death. The penalty fell on Henry of Ofterdingen ; in his despair he begged the Countess to gain him a respite so that he could go for his master, Klingsor. Her prayer was granted, and in the end Henry of Ofterdingen saved his head, though the legend says that Satan aided him. This story is without doubt founded on truth, but has much of fancy mingled with it.

The next remarkable story connected with Wartburg is the residence here of St. Elizabeth of Hungary, as she is called. This wonderful woman was the daughter of the King of Hungary, and when four years old she was betrothed to Prince Louis, son of Count Hermann, mentioned above. At this tender age she was given to his family. Her life at Wartburg was very remarkable, and I advise you to read about it, for it is too long to be given here. At last, her husband having died in Jerusalem, where he had gone with the Crusaders, his brother Henry drove her out with her children to seek a home where she could. She suffered much, and supported herself by spinning wool. But when the knights who had gone with her husband returned, they obliged Henry to give the son of Elizabeth his rights. She received the city of Marburg as her dower, but she did not live long. Miraculous things are told of her, and she is often represented by painters and sculptors.

Again, Wartburg was the residence of a remarkable

person ; for Luther dwelt there after escaping from the Diet at Worms.   He was called Ritter George, and the room where he wrote and spent much of his time is shown to travellers who visit the castle.

We come back now to Italy, the country we left when we passed from the Romanesque to Gothic architecture.   In the north of Italy where the Gothic order had prevailed after the eleventh century, it had been modified by the Romanesque influences and Roman traditions, in some such degree as the Moors had influenced the Gothic order in Spain.   But, on the whole, the mediæval buildings of Northern Italy were Gothic in style.

Rome, as we said, was individual, and her art remained Roman or Romanesque up to the date of the Renaissance.   In Southern Italy, as we shall see, the architecture was of the Byzantine order.

Among the most interesting edifices of the Middle Ages are the Italian towers. They were frequently quite

FIG. 90.—TOWER OF CREMONA.

separate from the churches and were built for various purposes. Some of them were bell towers, and such a tower
was called a *campanile*. Others were in some way associated
with the civic power of the cities which built them ; but the
largest number were for religious uses.

The *campanile* is always square at the bottom and for
some distance up, and then is frequently changed to an
octagonal or circular form and finished with a slender spire
or ornamental design.

Fig. 90 shows one of the finest square towers in
all Italy. It was built in 1296 to commemorate a peace
after a long war. It is three hundred and ninety-six feet
high. It has little beauty in the lower two thirds ; above
that it is more pleasing, but the two parts do not look as if
they belonged together. The tower of Italy, however,
which is most beloved and most famous is that of Giotto,
beside the cathedral of Florence. (See Fig. 102.)

Another striking feature of Gothic art in Northern Italy
is seen in the porches attached to the churches. They are
commonly on the side, and as they were usually added after
the rest of the church was finished, and frequently do not
correspond to the rest in style, they look as if they were
parts of some other churches and had come on a visit to
those beside which they stand. In Italy the main portion
of these porches always rested on lions.

A porch at Bergamo is one of the finest, and certainly
its details are exquisite, and the whole structure is beautiful
when it is considered separately ; but as a part of the church
it loses its effect, and seems to be pushed against it as a
chair is placed beside the wall of a room.

Some of the mediæval town-halls are still well preserved,
and a few of them are truly beautiful. Perhaps the Broletto
at Como is as fine a remnant of civic architecture as exists
in Northern Italy. It is not very large and is faced with
party-colored marbles.

FIG. 91.—ST. MARK'S CATHEDRAL. *Venice*

The architecture of Venice and the Venetian Province must be treated almost as if it were outside of Italy, because it differs so much from that of other portions of that country. During the Middle Ages it was the most prosperous portion of Italy. Its architecture was influenced by the Byzantine and Saracenic orders, but is not like them ; neither is it like that of Northern Italy ; in fact, it is Venetian, being Gothic in principle, but treated with Eastern feeling and decorated in Oriental taste ; and this was quite natural since the Venetians had extensive traffic and intercourse with the nations of the East.

There are few places in the world, of no greater extent, about which so many interesting associations cluster as about the Piazza of St. Mark's in Venice. On one side stands the great basilica, and not far away are the *campanile* and the clock-tower ; the ancient Doge's Palace, and the beautiful Library of St. Mark, of later date, are near by, with their treasures of art and literature to increase the value of the whole. It is a spot dear to all, and especially so to English-speaking people, since the poetry of Shakespeare has given them a reason for personal interest in it under all its varying aspects. At some hours of the day St. Mark's seems as if it were the very centre of the earth, to which men of all nations are hastening ; again this bustle dies away, and one could fancy it to be forgotten and deserted of all mankind, though its silence is eloquent in its power to recall the great events of the Venice of the past. (See Figs. 91, 105, and 106.)

St. Mark's Basilica is called Byzantine in its order, and in a general way the term is applicable to it ; but on careful examination there are so many differences between it and a purely Byzantine church that it would be more properly described by the name Italian or Venetian Byzantine. Its five domes were added to its original form late in the Middle Ages, and though there are many Eastern mosques with this number, they are not arranged like those of St. Mark's, and so have quite a different appearance. The portico with its five entrances is not European in form, but the details of these deep recesses are more like the Norman architecture than like anything Byzantine.

It is scarcely profitable to carry this examination farther, for, in a word, the whole effect of St. Mark's is very impressive from the exterior, and the interior is so beautiful in its subdued light and shadow that one is satisfied to enjoy it without criticising it. and many critics consider it one of the finest interiors of Western Europe.

FIG. 92.—SECTION OF SAN MINIATO. *Near Florence.*

The same difficulty which one finds in defining or classing the architecture of Venice is met in that of Southern Italy, which is Byzantine and not Byzantine, but, in fact, is that order so changed that the name of Byzantine-Romanesque seems better suited to it than any other term could be. We shall mention but a single example of this order, and pass to the true Byzantine style.

The church of San Miniato, which overlooks the city of Florence, was built in 1013, and is one of the most perfect as well as one of the earliest of the churches of the Byzantine-Romanesque order in Italy. It is not large, but the

FIG. 93.—SAN GIOVANNI DEGLI EREMITI. *Palermo.*

proportions are so good as to make it very pleasing ; the
pillars are so nearly classic in design that they were probably taken from some earlier building, and the effect of
colored panelling both within and without is very satisfactory to the eye.   (See Fig. 92.)

There arose in Sicily in the eleventh century, and after
the Norman Conquest, a remarkable style of architecture.
It belongs to Christian art because it was used by Christians
to construct places of Christian worship ; but, in truth, it
was a combination of Greek spirit with Roman form and
Saracenic ornament.   It makes an interesting episode in the
study of architecture.   I shall give one picture of a church

built by King Roger for Christian use as late as 1132, which, except for the tower, might well be mistaken for a purely Oriental edifice (Fig. 93).

BYZANTINE ARCHITECTURE.

This term strictly belongs to the order which arose in the East after Constantinople was made the Roman capital. It is especially the order of the Greek Church as contrasted with the Latin or Roman Church. It would make all architectural writing and talking much clearer if this fact were kept in mind ; but, unfortunately, wherever some special bit of carving in an Oriental design or a little colored decoration is used—as is frequently done in the modern composite styles of building—the term Byzantine is carelessly applied, until it is difficult for one not learned in architecture to discover what the Byzantine order is, or where it belongs.

We have spoken of its influence and partial use in Italy. Now we will consider it in its home and its purity. Before the time of Constantine the architecture used at Rome was employed at Jerusalem, Constantinople, and other Eastern cities which were under Roman rule and influence. Between the time of Constantine and the death of Justinian, in A.D. 565, the true ancient Byzantine order was developed. The church of St. Sophia, at Constantinople, was the greatest and the last product of the pure old Byzantine style.

From that time the order employed may be called the Neo-Byzantine. This was a decline of art as much as the history of Greece and the Eastern Empire during the same period (about 600 to 1453) was the history of the decline and extinction of a power that had once been as great among governments as St. Sophia (Fig. 94) was among churches.

The chief characteristic of Byzantine architecture is the use of the dome, which is the most important part of its

FIG. 94.—CHURCH OF ST. SOPHIA. Constantinople. Exterior View

design. A grand central dome rises over the principal portion of the edifice, and just as in other orders courts and colonnades were added to the simpler basilica form in the ground plan of the churches, so in the Byzantine order lesser domes and cupolas were added above until almost any number of them was admissible, and they were placed with little attention to regularity or symmetry of arrangement.

As domes were the chief exterior feature, so the profuse ornamentation was most noticeable in the interior. The walls were richly decorated with variegated marbles;

FIG. 95.—LOWER ORDER OF ST. SOPHIA.

the vaulted ceilings of the domes and niches were lined with brilliant mosaics; the columns, friezes, cornices, door and window-frames, and the railings to galleries were of marbles, and entirely covered with ornamental designs (Figs. 95 and 96).

The historian Gibbon describes the building of St. Sophia and its decorations. He tells us that the emperor went daily, clad in a linen tunic, to oversee the work. The architect was named Anthemius; he employed ten thousand workmen, and they were all paid each evening. When it was completed and Justinian was present at its consecration, he exclaimed, "Glory be to God, who hath thought

me worthy to accomplish so great a work ; I have van-
quished thee, O Solomon !"

Paul Silentiarius was a poet ; he saw St. Sophia in all its
glory and describes it with enthusiasm. It was very rich in
variegated marbles. He mentions the following : 1. *The
Carystian*, pale with iron veins. 2. *The Phrygian*, two
sorts, both of a rosy hue ; one with a white shade, the
other purple with silver flowers. 3. *The Porphyry of
Egypt*, with small stars. 4. *The green marble of Laconia*.
5. *The Carian*, from Mount Iassis, with oblique veins, white
and red. 6. *The Lydian*, pale,
with a red flower. 7. *The
African or Mauritanian*, of a
gold or saffron hue. 8. *The
Celtic*, black, with white veins.
9. *The Bosphoric*, white, with
black edges. There were also
the *Proconnesian*, which made
the pavement ; and the *Thes-
salian* and *Molossian* in different parts.

This array of marbles was
made even more effective by
the beautiful columns brought
from older temples. The mo-
saics were rich in color, and numerous, and many parts of
the church were covered with gold, so that the effect was
dazzling.

FIG. 96.—UPPER ORDER OF
ST. SOPHIA.

Those objects that were most sacred were of solid gold
and silver, while such as were less important were only
covered with gold-leaf. In the sanctuary there was altogether forty thousand pounds of silver ; the vases and vessels used about the altar were of pure gold and studded
with gems. Its whole cost was almost beyond belief. At
the close of his description Gibbon says : " A magnificent

Fig. 97.—Interior View of Church of St. Sophia.

temple is a laudable monument of taste and religion, and
the enthusiast who entered the dome of St. Sophia might
be tempted to suppose that it was the residence or even the
workmanship of the Deity. Yet how dull is the artifice,
how insignificant is the labor, if it be compared with the
formation of the vilest insect that crawls upon the surface
of the temple !''

Of course, individual taste must largely influence the
opinion regarding the beauty of any work of art, but to me
St. Sophia, which is the chief example of Byzantine architecture, is far less beautiful and less grand than the finest
Gothic cathedrals. Comparatively little attention was paid
to the elegance and decoration of the exterior in the
Eastern edifices, while the interiors, in spite of all their
riches, have a flat and unrelieved effect. Probably the
chief reason for this is that color is substituted for relief—
that is to say, in Gothic architecture heavy mouldings and
panellings, though of the same color as the walls themselves, yet produce a marvellous effect of light and shadow,
and even lend an element of perspective to various parts of
the building. In the place of these mouldings flat bands of
color are often used in the Byzantine order, and the whole
result is much weakened, though a certain gorgeousness
comes from the color. Another cause of disappointment in
St. Sophia is the absence of painted glass. At the same
time, and in spite of these defects, St. Sophia is grand and
beautiful—but not solemn and impressive in comparison
with the dim cathedral aisles of many Gothic churches in
other parts of the world. (See Fig. 97.)

The Romanesque and Byzantine styles came at last to
be so mingled that it would be folly to attempt to separate'
their influence, but the Byzantine had much more originality, and left a far wider mark.

Among the most noted examples of the latter style,
beside St. Sophia and St. Mark's, are the church of St.

Vitale at Ravenna, the cathedral at Aix-la-Chapelle, supposed to have been built by Charlemagne about 800 A.D., and the church of the Mother of God at Constantinople.

SARACENIC ARCHITECTURE.

In speaking of Saracenic architecture I will first explain that it is one with the Moresco or Moorish order of which I spoke in connection with Spain. The only difference is, that the earliest Mohammedan conquerors of Spain are said to have come from ancient Mauri or Mauritania and were called Moors, while the name of *Saraceni*, which means "the Easterns," was also given to them. Thus the Mohammedan architecture in Spain is called both Moresco, or Moorish, and Saracenic. Again, it is also called Arabian, but I think this is the least correct, since the Easterns who went to Spain were not so universally Arabian as to warrant this name. When we speak of Moresco or Moorish architecture we speak of Spain; but the term Saracenic is used for Mohammedan architecture in all countries where it is found, and is a just term, for they are Eastern or Oriental lands.

In absolute fact, Saracenic architecture is that of the followers of "the Prophet," as Mohammed is called, and would be more suitably named if it were called Mohammedan architecture, or the architecture of Islam.

Mohammed was born at Mecca A.D. 570, but it was not until 611 that he was commissioned, as he believed, to build up a new faith and a new church. At first his followers were so few and so mingled with other sects and tribes in their outward life that they had no distinctive art. It was not until A.D. 876, when the ruler Ibn-Touloun commenced his splendid mosque at Cairo, that the Mohammedans could claim any architecture as their own. It is very interesting to know that there were pointed arches in this mosque,

probably two centuries, at least, earlier than they were used
in England, for it is generally believed that they were first
used there in the rebuilding of Canterbury Cathedral after
it was burned in 1174. When, however, the Saracenic
order was fully established it was so individual and so different from all other architecture that there is no mistaking it
for that of any other religion or nation than that of
Mohammed and his followers.

FIG. 98.—MOSQUE OF KAITBEY.

FIG. 99.—THE CALL TO PRAYER.

The picture of the mosque of Kaitbey shows one of the finest and most elegant mosques of the East. It is just outside the walls of Cairo, and is quite modern, having been built in 1463. This view of it gives an excellent idea of the appearance of a fine mosque and shows the minaret or tower, which is so important in a mosque, to good advantage (Fig. 98).

These minarets are constantly used for the many calls to prayer which are made throughout the day and night. The person who makes these calls is styled " the Muezzin," and is usually blind. Several times during the day he ascends the minaret and calls out in a loud and melodious tone, "God is most great ; there is no God but Allah, and I testify that Mohammed is Allah's prophet ! Come to prayer ! Come to security ! Prayer is better than sleep !" This is several times repeated and is called the *Adan*.

The form of words used for the night varies a little, ending, "There is no God but Allah. He has no companion ! To Him belongs dominion, etc.;" this is called the *Ula*. The call made an hour before day is the *Ebcd*, and praises the perfection of God. When one is sleeping near enough to a minaret to hear the muezzin's voice it is a pleasant sound and helps one to realize that the care of God is ever about him ; the clear, Christian bell can be heard by more people, and this was originally intended as a call to prayer. (See Fig. 99.)

The principal homes of Saracenic architecture are Syria, Egypt, Mecca, Barbary, Spain, Sicily, Turkey, Persia, and India. There are many very interesting mosques and minarets that might be mentioned had we space, but I can speak only of the mosque of Cordova, which is universally admitted to be the finest Saracenic edifice in the world (Fig. 100), and shall quote a part of the interesting description of it given by De Amicis in his delightful book called "Spain and the Spaniards."

This mosque was commenced by the Caliph Abd-er-Rahman in 786, and was completed by his son Heshâm, who died 796. The great Caliph declared that he would build a mosque which should exceed all others in the world and be the Mecca of the West. De Amicis, after describing the garden which surrounds the mosque, enters, and then goes on as follows : " Imagine a forest, fancy yourself in the thickest portion of it, and that you can see nothing but the trunks of trees. So, in this mosque, on whatever side you look, the eye loses itself among the columns. It is a forest of marble whose confines one cannot discover. You follow with your eye, one by one, the very long rows of columns that interlace at every step with numberless other rows, and you reach a semi-obscure background, in which other columns still seem to be gleaming. There are nineteen naves, which extend in every direction, traversed by thirty-three others, supported (among them all) by more

Fig. 100.—Exterior of the Sanctuary in the Mosque of Cordova.

than nine hundred columns of porphyry, jasper, breccia, and marbles of every color. Each column upholds a small pilaster, and between them runs an arch (see plate above), and a second one extends from pilaster to pilaster, the latter placed above the former, and both of them in the shape of a horseshoe ; so that, in imagining the columns to be the trunks of so many trees, the arches represent the branches, and the similitude of the mosque to a forest is complete. . . . How much variety there is in that edifice which at first sight seems so uniform ! The proportions of the columns, the designs of the capitals, the forms of the arches change, one might say, at every step. The majority of the columns are old, and were taken from the Arabs of Northern Spain, Gaul, and Roman Africa, and some are said to have belonged to a temple of Janus, on the ruins of which was built the church that the Arabs destroyed in order to erect the mosque. Above several of the capitals one can still see traces of the crosses that were cut on them, which the Arabs broke with their chisels. . . . I stopped for a long time to look at the ceiling and walls of the principal chapel, the only part of the mosque that is quite intact. It is a dazzling gleam of crystals of a thousand colors, a network of arabesques, which puzzles the mind, and a complication of bas-reliefs, gildings, ornaments, minutiæ of design and coloring, of a delicacy, grace, and perfection sufficient to drive the most patient painter distracted. . . . You might turn a hundred times to look at it, and it would only seem to you, in thinking it over, a mingling of blue, red, green, gilded and luminous points, or a very intricate embroidery changing continually, with the greatest rapidity, both design and coloring. Only from the fiery and indefatigable imagination of the Arabs could such a perfect miracle of art emanate. . . . Such is the mosque of to-day. But what must it have been in the time of the Arabs ? It was not surrounded by a wall, but open, so that one could catch a glimpse of the garden from every part of it ; and from the garden one could see to the end of the long naves, and the air was full of the fragrance of oranges and flowers. The columns which now number less than a thousand were then fourteen hundred ; the ceiling was of cedar-wood and larch, sculptured and enamelled in the finest manner ; the walls were trimmed with marble ; the light of eight hundred lamps, filled with perfumed oil, made all the crystals in the mosaics gleam, and produced on the pavements, arches, and walls a marvellous play of color and reflection. 'A sea of splendors,' sang a poet, ' filled this mysterious recess ; the ambient air was impregnated with aromas and harmonies, and the thoughts of the faithful wandered and lost themselves in the labyrinth of columns which gleamed like lances in the sun.' "

The famous palace of the Alhambra is so well known that I cannot leave this part of our subject without one picture and one bit of description of it from the same author, De Amicis.

The Alhambra was built about four centuries ago, and the wall which inclosed it was four thousand feet long by twenty-two hundred feet wide. Within this there were gardens, fountains, kiosks, and many beautiful, fanciful structures, all of which doubtless cost as much as the more necessary parts of the edifice. The roofs of the different parts of the palace were supported by forty-three hundred columns of precious marbles ; eleven hundred and seventy-two of these were presented to Abd-er-Rahman (for he was also the founder of the Alhambra) by sovereigns of other countries, or else brought by him from distant shores for the decoration of this splendid, fairy-like place. All the pavements were of beautiful marbles ; the walls, too, were of the same material, with friezes arranged in splendid colors ; the ceilings were of deep blue color, with figures in gilding and interlacing designs running over all. In truth, nothing that could be imagined or wealth buy to make this palace beautiful was left out ; and yet we are told that the palace of Zahra which was destroyed was still finer. All this leads one to almost believe that the " Arabian Nights" are no fanciful tales, but quite as true as many more serious sounding stories.

The Court of the Lions is called " the gem of Arabian art in Spain," and of this our author says : " It is a forest of columns, a mingling of arches and embroideries, an indefinable elegance, an indescribable delicacy, a prodigious richness, a something light, transparent, and undulating like a great pavilion of lace ; with almost the appearance of a building which must dissolve at a breath ; a variety of lights, views, mysterious darkness, a confusion, a capricious disorder of little things, the majesty of a palace, the gayety of a kiosk, an amorous grace, an extravagance, a delirium, the fancy of an imaginative child, the dream of an angel, a madness, a nameless something—such is the first effect produced by the Court of the Lions !'' (Fig. 101.)

This court is not large ; the ceiling is high, and a light portico runs round it upheld by white marble columns in clusters of two, three, or more, so arranged as to resemble trees coming up from the ground. Above the columns the designs almost resemble curtains, and there are little graceful suggestions like ribbons and waving flowers. " From the middle of the shortest sides advance two groups of columns, which form two species of square temples of nine arches each (see cut) surmounted by as many colored cupolas. The walls of these little temples and the exterior of the portico are a real lace-work of stucco, embroideries, and hems, cut and pierced from one side to the other, and as transparent as net-work, changing in design at every step. Sometimes they end in points, in crimps, in festoons, sometimes in ribbons waving round the arches, in kinds of stalactites, fringes, trinkets, and bows which seem to move and mingle with each other at the slightest breath of air. Large Arabic inscriptions run along the four walls, over the arches, around the capitals, and on the walls of the little temples. In the centre of the court rises a great marble basin, upheld by twelve lions (see cut), and surrounded by a little paved canal. . . . At every step one takes in the court that forest of columns seems to move and change place, to form again in another way ; behind one column, which seems alone, two, three, or a row will spring out ;

FIG. 101.—COURT OF THE LIONS ALHAMBRA.

others separate, unite, and separate again. . . . We remained for more than an hour in the court, and it passed
like a flash ; I, too, did what almost all people do, be they
Spanish or strangers, men or women, poets or not. I ran
my hand along the walls, touched all the little columns,
and passed my two hands around them, one by one, as
around the waist of a child ; I hid among them, counted
them, looked at them on a hundred sides, crossed the court
in a hundred ways, tried if it were true that in saying a
word, *sotto voce*, into the mouth of one lion, one could hear
it distinctly from the mouths of all the others ; I looked on
the marbles for the spots of blood of poetic legends, and
wearied both brain and eye over the arabesques. . . . In
all my life I have never thought, nor said, nor shall I say,
so many foolish, stupid, pretty, senseless things as I said and
thought in that hour."

The study of Saracenic architecture in Turkey, Persia,
and India is very interesting, but our space warns us that
we must hasten to leave this dreamy, fairy-like part of our
subject and come down to later times and more realistic
matters.

CHAPTER III.

ALL Architecture since the time of the Renaissance is called Modern Architecture ; this term, therefore, embraces all edifices erected during nearly four centuries.

When I first spoke of Architecture I said that it was a constructive art, and not imitative like Painting and Sculpture. In its earlier history this was true, but the time came when it also became an imitative art and had no true or original style. The Gothic order was the last distinct order which arose, and since its decline, at the beginning of the Renaissance, all architecture has been an imitation because it is a reproduction of what existed before ; at times some one of the older orders has been in favor and closely imitated, and again, parts of several orders are combined in one edifice. Since the time of the Reformation it has been true, almost without exception, that every building of any importance has been copied from something belonging to a country and a people foreign to the land in which it was erected.

When the revival of Classic Literature began, Rome was the first to feel its influence. It was welcomed there with open arms, just as we might receive the early history and literature of our country if it had all been lost and was found again ; for this was precisely what it meant to the

Romans, when, after the Dark Ages, the works of Livy, Tacitus, and Cæsar were in their hands, and they read of the history, art, and literature of their past. They were enthusiastic, and their feeling soon spread over all Italy.

France was the next to adopt the newly-revived ideas, for that country looked to Rome as the source of true religion, and a model in all things. Spain was then in an unsettled state, and welcomed the revival of classic art as heartily as it had already embraced the Church of Rome.

In Germany the love of the classics was enthusiastic, but that nation was more taken up with literature and slower in adopting the revival of the arts than were the more southern peoples, and the fifteenth and sixteenth centuries are a barren period in the history of German architecture.

In England, too, the Renaissance made slow progress. It was not until the time of Charles I. that any influence was felt in Great Britain from the revival of classic taste which was so well established on the Continent.

As it is true that no new order of Architecture has arisen since the time of those of which I have already told you, I shall try to make you understand something of Modern Architecture by speaking of certain important edifices in one country and another, with no attempt at any more detailed explanation of it.

ITALY.

We cannot say that the art of the Renaissance originated in one city or another, because the movement in the revival of art was so general throughout Italy ; but Florence has a strong claim to our first consideration from the fact that Filippo Brunelleschi was a Florentine and did his greatest work in his native city, and on account of it has been called " the father of the Art of the Renaissance." He was born in 1377, and from his early boyhood was inclined to be an architect. The cathedral of Florence (Fig. 102), which is

FIG. 102.—THE CATHEDRAL OF FLORENCE AND GIOTTO'S CAMPANILE.

also called the church of Sta. Maria del Fiore, had been
built long before, but had never been finished by a roof or
dome.

Brunelleschi was possessed with but one desire, which
was to complete this cathedral. He went to Rome and
diligently studied the remains of classic art which he found
there, and especially the dome of the Pantheon. Return-
ing to Florence he took measures to bring his plans before
the superintendents of the cathedral works ; he was
ridiculed and discouraged on every hand, but he never gave
up his hopes nor lessened his study of the ways and means
by which the dome could be built. Thus many weary years
passed by ; Brunelleschi made drawings in secret, and from
these he constructed models in order to convince himself of
what he could do.

At last those who had authority in the matter were
ready to act, and a convention was called, before which the
architects of different nations appeared and were requested
to explain their theories of what could be done to cover the
cathedral. Many artists were assembled and various plans
were shown, but after all had been examined the work was
given to Brunelleschi, and he was happy in finding that the
years he had devoted to the study of the dome had not
been spent in vain.

It was on this occasion that Brunelleschi refused to
show his models, and when the other architects blamed him
for this he asked that some eggs should be brought, and
proposed that he who could make an egg stand upright on
a smooth piece of marble should be the builder of the
dome. The others tried to do this and failed ; at last
Brunelleschi brought his egg down on the marble with a
sharp tap and left it standing erect. Then all exclaimed,
" Oh, we could have done that if we had known that was
the way," to which Brunelleschi replied, " So you could
have built a dome if I had shown you my models."

FIG. 103.—VIEW OF ST. PETER'S, Rome.

This story is often told of Columbus, but as Brunelleschi
was much older than Columbus, and the fact is related by
Florentine writers of his time, it is probable that Columbus
had heard of it from the geographer Toscanelli, who was a
great admirer of Brunelleschi and a friend of Columbus
also. In building the dome, Brunelleschi encountered great
difficulties, but he lived to be assured of his success, for at
his death, in 1444, it lacked but little of completion, and all
the parts essential to its perfection and durability were
finished.

This is the largest dome in the world, for though the
cross on the top of St. Peter's is farther from the ground
than that of Florence, the dome itself above the church is
not as large as the dome of Sta. Maria del Fiore.

This work made Brunelleschi's greatest fame, but he
was the architect of many other fine churches and of secular
buildings also ; among the last the Pitti Palace, in which is
the famous Pitti Gallery, is one of the most important.
When you go to Florence you will see a statue of Filippo
Brunelleschi, which is very interesting, on account of the
way in which it is represented and the position in which it
is placed.   It is on one side of the Piazza of the cathedral ;
he is calmly sitting there with a plan of the church spread
before him on his lap, while he lifts his head to look at the
great dome as it stands out against the sky, the realization
of all his thought and labor during so many years.

The church of St. Peter's at Rome, which is the largest
and most magnificent of all Christian temples, was begun
about 1450, and was not brought into its present form until
about 1661, or more than two centuries later (Fig. 103).

The history of its building is largely a story of conten-
tions and troubles between popes, architects, and artists of
different kinds.   As it now stands it is as much the work
of Michael Angelo as of any one man, but several other
architects left their imprint upon it, both before and after

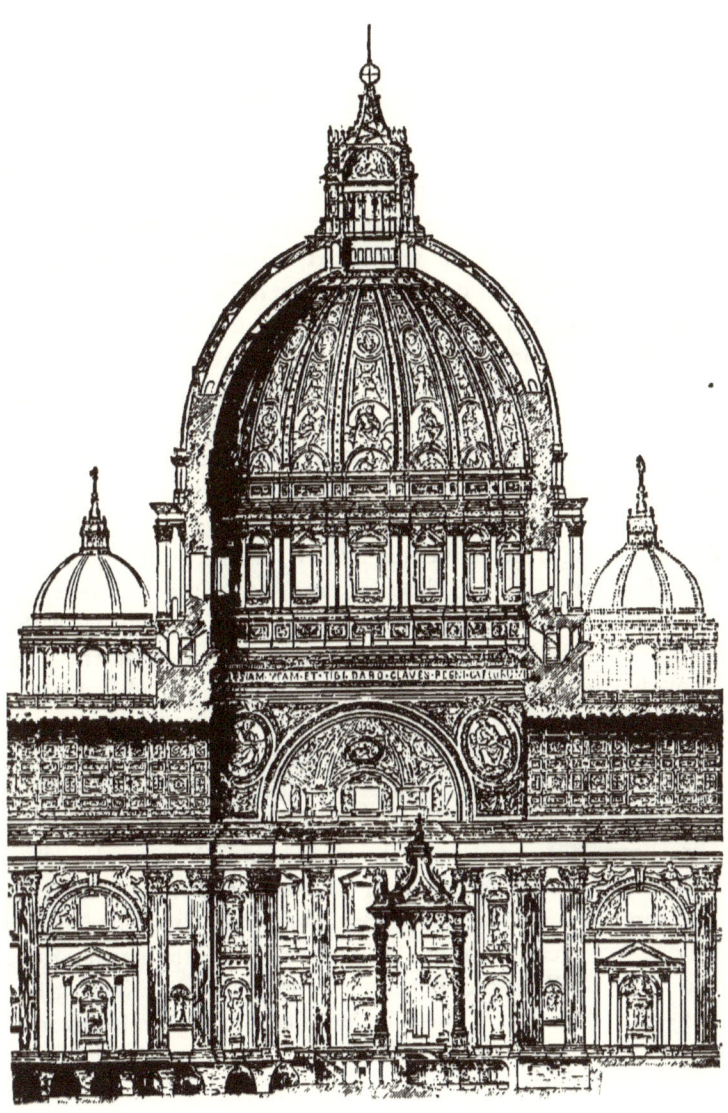

FIG. 104.—SECTION OF ST. PETER'S.

his time ; and all who aided in its construction were eminent men, in their way.   Michael Angelo was in his seventy-
second year when he took up the task of completing St.
Peter's.   Bramante, Raphael, and Peruzzi had preceded
him as architects of the church ; Michael Angelo designed
the dome, and when he was ninety it was nearly finished ;
the models for its completion which he made were not
followed after his death ; his plan would have made the
church more harmonious with the dome, in size, than it
now is.   Money was sent in large sums, from all Europe,
to carry on this work ; the finest materials were used in
building it, and the most gifted artists were employed in its
decoration ; it is now the vast home of multitudes of treasures.   "I have hung the Pantheon in the air !" Michael
Angelo is said to have exclaimed, while looking at the
splendid dome of St. Peter's ; and no dome in the world
has a more imposing effect, although its harmony with the
rest of the building is injured by the change of the plan
from that of a Greek cross which was made after his death.*

In spite of all this the critics of architecture are never
weary of pointing out the defects of St. Peter's ; but to
those who cannot apply to it the test of strictly scientific
rules, its interior is sublime in its effect, and has few rivals
—perhaps but one—in the world, and that is the great
Hypostyle Hall at Karnak, of which we spoke when writing
of Egyptian architecture.   But even here the difference is
almost too great to admit of comparison ; the spirit of the
two is so unlike—St. Peter's is complete and Karnak is a
ruin—so, after all, it must be admitted that the interior of
St. Peter's is superior to all other edifices of which we
know (Fig. 104).

* The interior diameter of the dome of St. Peter's is one hundred and
thirty-nine feet ; that of St. Sophia, one hundred and fifteen feet, and that of
Sta. Maria del Fiore, at Florence, one hundred and thirty-eight feet, six
inches.

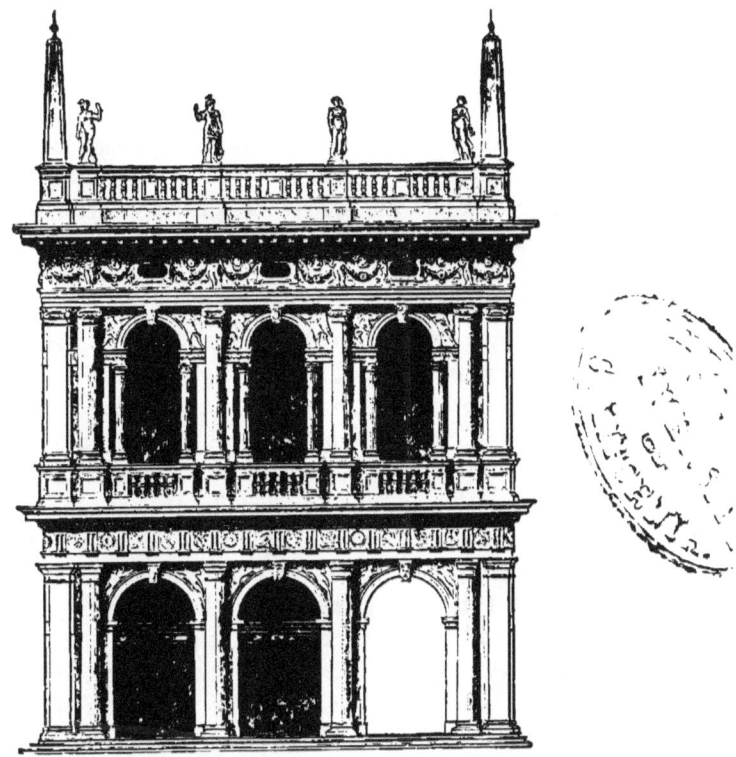

FIG. 105.—EAST ELEVATION OF LIBRARY OF ST. MARK.   *Venice*

From the time of the beginning of the Renaissance, about 1420, to about 1630, the architecture of Venice was going through a change, and finally reached such perfection that during the next half century the most magnificent style of architecture prevailed which has ever been known there. We mean to say that the whole effect was the grandest, for, while it is true that the edifices of that time are stately and striking in their appearance, it is equally true that their form and ornamentation are not as much in keeping with their use as they had been in older edifices.

Sansovino, who lived from 1479 to 1570, was an important architect and had great influence upon modern Venetian
architecture. His masterpiece was the Library of St.
Mark, of which the preceding cut gives one end (Fig. 105).
It is a very beautiful structure, and is made more interesting from the fact that it stands directly opposite to the
Doge's Palace, and in the midst of all the interest which
centres about the Piazza of St. Mark.

The Ducal Palace at Venice is called by John Ruskin,
the great English critic, " the central edifice of the world."
It is divided into three stories, of which the uppermost
occupies rather more than half the height of the building.
The two lower stories are arcades of low, pointed arches,
supported on pillars, the one beneath being bolder and
heavier in character than the second. The capitals of the
columns are greatly varied, no two in the upper arcade
being exactly alike. Above the arches of the middle story
was a row of open-work spaces, of the form called quatrefoil ; while the third story is faced with alternating blocks
of rose-colored and white marble, and is pierced with a few
large pointed windows. The whole front, or façade, is
crowned by an open parapet made up of blocks of stone
carved into lily-like forms alternating with lance-shaped
leaves. The whole effect is one of great richness and
beauty, especially since time has mellowed its color,
and softened without destroying the whiteness of its
marbles (Fig. 106).

During the time of the Renaissance there were churches,
palaces, museums, hospitals, and other large buildings
erected in all the important cities of Italy. There are but
few of these which have such special features as entitle
them to be selected for description here. The reason for
this has been given already—viz. : there was nothing new
in them ; they were all repetitions of what has been described in one form or another. Perhaps the next cut gives

FIG. 106.—THE DOGE'S PALACE. *Venice.*

FIG. 107.—GREAT COURT OF THE HOSPITAL OF MILAN.

as good an example of secular architecture in this age as
any that could be selected (Fig. 107).

Indeed, it is one of the most remarkable buildings of its
class in any age. It was commenced by Francesco Sforza
and his wife, Bianca, in 1456. They died long before its
completion, and one part and another have been changed
from time to time, but its great court, which was designed
by Bramante, still remains, the finest thing of its kind in all
Italy.

I shall now leave Italy with saying that the early days
of the Renaissance were the best days of Italian Architecture, and, indeed, of Italian Art. The period made sacred
by the genius and works of Michael Angelo, Bramante,

Sangallo, Leonardo da Vinci, and Raphael was a golden era, and still sheds its lustre over the land of their nativity. These artists followed the highest ideal of Art, and their errors were superior to the so-called successes of less gifted men.

The Italian Art of the fifteenth century was individual and grand ; in the sixteenth century it became formal and elegant ; in the seventeenth century it was bizarre, over-ornamented, and uncertain in its aim and execution ; since then it has been comparatively unimportant, and its architecture scarcely merits censure, and certainly cannot be praised.

SPAIN.

From the time of the fall of Granada, in 1492 to 1558, Spain was the leading nation of Europe. The whole country had been united under Ferdinand and Isabella, and their reign was a glorious period for their country. The importance of the nation was increased by the discovery of the New World, and so many great men were in her councils that her eminence was sure, and almost undisputed. Thus it followed that during the first half of the sixteenth century the Architecture of Spain gave expression to the spirit by which the nation was then animated.

This did not long continue, however, for the iron, practical rule of Philip II. crushed out enthusiasm and was fatal to artistic inspiration. This sovereign desired only to extend his kingdom ; the priests, who acquired almost limitless power under his reign, aimed only to strengthen their authority, while the people were wildly pursuing riches in the New World which opened up to them a vast and attractive field. Thus no place or time was left to the cultivation of Art, and the only noteworthy period of Spanish Architecture since the beginning of the Renaissance was the sixty years which we have mentioned.

The Modern Architecture of Spain has been divided into three eras, each of which was distinguished by its own style. The first extends from the beginning of the Renaissance down to that of the abdication of the great Emperor Charles V. in 1555 ; the manner of this period is called Platerisco, or the silversmith's style, on account of the vast amount of fine, filigree ornament which was used. The second period is from the above date to about 1650, and its art is called the Græco-Roman style because it is an attempt to revive the Classic Art of the ancient Greeks and Romans. The third period comes from 1650 to about a century later, and the Spaniards call its manner the Churrigueresque, which difficult name they take from that of Josef de Churriguera, the architect who invented this style. Since 1750 we may almost say that no such thing as Spanish Architecture has existed.

The cathedrals of Granada, Jaen, and Valladolid, and the churches of Malaga and Segovia, with many other ecclesiastical edifices, are among the chief monuments of Spanish Renaissance Architecture, but we shall pass on to a little later period and speak of but one great achievement, the famous Escurial, which is of much historic interest.

This combination of basilica, palace, monastery, and college was begun in 1563 by Philip II., in accordance with a vow which he made to St. Lawrence at the battle of St. Quentin. This battle was fought in 1557 under the walls of the French town of St. Quentin, by the French and the Spaniards, and the latter were completely victorious.

This cut gives an idea of how grand and impressive this collection of walls, towers, courts, and edifices must be, all crowned with the dome of the basilica. It is almost like a city by itself, and all who visit it agree that it is a gloomy and depressing place in spite of its grandeur (Fig. 108).

The front has three imposing entrances, with towers at the corner angles. Within the inclosure are a college,

FIG. 108.—THE ESCURIAL. *Near Madrid.*

monastery, palace with state apartments, the church,
numerous courts, gardens, and fountains. The front is
injured by the great number of small windows, which divide
it into such numberless sections as to become very tiresome
to the eye, while they take away the noble elegance of
larger spaces and the air of repose which such spaces give.
The angle towers are not as rich in effect as they should be,
and the side walls have been compared to those of a Man-
chester cotton-mill ; thus the exterior, which is effective
from its size and general air, has not the beauty of detail
which satisfies a close observer

The effect of the interior, as one goes in by the central entrance, is all that can be desired. The court leads directly to the square before the church ; as one passes to it he has the college on one side, the monastery on the other, farther on the palace, with the whole culminating in the grand state apartments and the basilica. The various courts are striking in their arrangement, and the church with its dome and towers gives a supreme glory to the whole. Gardens, fountains, and many other fine objects add their effect to the richness and beauty of the whole ; but all are insignificant beside the basilica, which merits a place in the foremost rank of the churches of the Renaissance. Indeed, the Escurial is a marvellous place, and is often called " the eighth wonder of the world." The richest marbles, splendid pictures, and many magnificent objects help to make it one of the grandest works of modern architecture.

It is also true that it is one of the gloomiest places visited by travellers, and I shall quote a few lines from De Amicis to show the depressing effect which it has upon those who go there.

" The first feeling is that of sadness ; the whole building is of dirt-colored stone, and striped with white between the stones ; the roofs are covered with strips of lead. It looks like an edifice built of earth. The walls are very high and bare, and contain a great number of loopholes. One would call it a prison rather than a convent. . . . The locality, the forms, the colors, everything, in fact, seems to have been chosen by him who founded the edifice with the intention of offering to the eyes of men a sad and solemn spectacle. Before entering you have lost all your gayety ; you no longer smile, but think. You stop at the doors of the Escurial with a sort of trepidation, as at the gates of a deserted city ; it seems to you that, if the terrors of the Inquisition reigned in some corner of the world, they ought to reign among those walls. You would say that therein one might still see the last traces of it and hear its last echo. . . . The royal palace is superb, and it is better to see it before entering the convent and church, in order not to confuse the separate impressions produced by each. This palace occupies the northeast corner of the edifice. Several rooms are full of pictures, others are covered from floor to ceiling with tapestries, representing bull-fights, public balls, games, fêtes, and Spanish costumes, designed by Goya ; others are regally furnished and adorned ; the floor, the doors, and the windows are covered with marvellous inlaid work and stupendous gilding. But among all the rooms the most noteworthy is that of Philip II. ; it is rather a cell than a room, is bare and squalid, with an alcove which answers to the royal oratory of the church, so that, from the bed, by keeping the doors open, one can see the priest who is saying mass. Philip II. slept in that cell, had his last illness there, and there he died. One still sees some chairs used by him, two little stools upon which he rested the leg tormented with gout, and a writing-desk. The walls are white, the ceiling flat and without any ornament, and the floor of brick. . . In the court-yard of the kings you can form a first idea of the immense frame-work of the edifice. The court is inclosed by walls ; on the side opposite the doors is the façade of the church. On a spacious flight of steps there are six enormous Doric columns, each of which upholds a large pedestal, and every pedestal a statue. There are six colossal statues, by Battiste Monegro, representing Jehoshaphat, Ezekiel, David, Solomon, Joshua, and Manasseh. The court yard is paved, scattered with bunches of damp turf. The walls look like rocks cut in points ; everything is rigid, massive, and heavy, and presents the fantastic appearance of a Titanic edifice, hewn out of solid stone, and ready to defy the shocks of earth and the lightnings of heaven. There one begins to understand what the Escurial really is.

"One ascends the steps and enters the church. The interior is sad and bare. . . . Beside the high altar, sculptured and gilded in the Spanish style, in the inter-columns of the two royal oratories, one sees two groups of bronze statues kneeling, with their hands clasped toward the altar. On the right Charles V. and the Empress Isabella, and several princesses ; on the left, Philip II. with his wives. . . . In a corner, near a secret door, is the chair which Philip II. occupied. He received through that door letters and important messages, without being seen by the priests who were chanting in the choir. This church, which, in comparison with the entire building, seems very small, is nevertheless one of the largest in Spain, and although it appears so free from ornamentation, contains immense treasures of marble, gold, relics, and pictures, which the darkness in part conceals, and from which the sad appearance of the edifice distracts one's attention. . . . But every feeling sinks into that of sadness. The color of the stone, the gloomy light, and the profound silence which surrounds you, recall your mind incessantly to the vastitude, unknown recesses, and solitude of the building, and leave no room for the pleasure of admiration. The aspect of the church awakens in you an inexplicable feeling of inquietude. You would divine, were you not otherwise aware of it, that those walls are surrounded, for a great distance, by nothing but granite, darkness, and silence ; without seeing the enormous edifice, you feel it ; you feel that you are in the midst of an uninhabited city ; you would fain quicken your pace in order to see it rapidly, to free yourself from the weight of that mystery, and to seek, if they exist anywhere, bright light, noise, and life. . . . One goes to the convent, and here human imagination loses itself ; . . . you pass through a long subterranean corridor, so narrow that you can touch the walls with your elbows, low enough almost to hit the ceiling with your head, and as damp as a submarine

grotto ; you reach the end, turn, and you are in another
corridor. You go on, come to doors, look, and other corridors stretch away before you as far as the eye can reach
At the end of some you see a ray of light, at the end of
others an open door, through which you catch a glimpse of
a suite of rooms. . . . You look through a door and start
back alarmed ; at the end of that long corridor, into which
you have glanced, you have seen a man as motionless as a
spectre, who was looking at you. You proceed, and
emerge on a narrow court, inclosed by high walls, which is
gloomy, overgrown with weeds, and illumined by a faint
light which seems to fall from an unknown sun, like the
court of the witches described to us when we were children.
. . . You pass through other corridors, staircases, suites of
empty rooms, and narrow courts, and everywhere there is
granite, a pale light, and the silence of a tomb. For a
short time you think you would be able to retrace your
steps ; then your memory becomes confused, and you
remember nothing more ; you seem to have walked ten
miles, to have been in that labyrinth for a month, and not
to be able to get out of it. You come to a court and say,
' I have seen it already ! ' but you are mistaken ; it is
another. . . . You seem to be dreaming ; catch glimpses
of long frescoed walls ornamented with pictures, crucifixes,
and inscriptions ; you see and forget ; and ask yourself,
' Where am I ? ' . . . On you go from corridor to corridor,
court to court ; you look ahead with suspicion ; almost
expect to see suddenly, at the turning of a corner, a row of
skeleton monks, with their hoods drawn over their eyes and
their arms folded ; you think of Philip II., and seem to
hear his retreating step through dark hallways ; you remember all that you have read of him, of his treasures, the
Inquisition, and all becomes clear to your mind's eye ; you
understand everything for the first time ; the Escurial *is*
Philip II., he is still there, alive and frightful, and with

him the image of his terrible God. . . . The Escurial surrounds, holds, and overwhelms you ; the cold of its stones
penetrates to your marrow ; the sadness of its sepulchral
labyrinths invades your soul ; if you are with a friend you
say, ' Let us leave ; ' if you were alone you would take to
flight. At last you mount a staircase, enter a room, go to
the window, and salute with a burst of gratitude the mountains, sun, freedom, and the great and beneficent God who
loves and pardons. What a long breath one draws at that
window !

" An illustrious traveller said that after having passed a
day in the convent of the Escurial, one ought to feel happy
throughout one's life, in simply thinking that one might
still be among those walls, but is no longer there. This is
almost true. Even at the present day, after so great a lapse
of time, on rainy days, when I am sad, I think of the
Escurial, then look at the walls of my room, and rejoice !"

During the sixteenth century there were many palaces
erected in Spain, but nothing can be added to the impressions you will get from the descriptions we have quoted of
the cheerful, gay Alhambra, and the gloomy, sad Escurial.

The domestic architecture of Spain is unattractive.
There are no fine *châteaux*, as in France, or elegant parks,
as in England. Ford compares the front of the residence
of the Duke of Medina to " ten Baker-street houses put
together," and this is true of many so-called palaces. This
state of modern Spanish architecture is fully accounted for
by the following quotation from Fergusson, the learned
writer on architecture :

" On the whole, perhaps, we should not be far wrong in
assuming that the Spaniards are among the least artistic
people in Europe. Great things have been done in their
country by foreigners, and they themselves have done
creditable things in periods of great excitement, and under
the pressure of foreign example ; but in themselves they

seem to have no innate love of Art, no real appreciation for its beauties, and, when left to themselves, they care little for the expression of beauty in any of the forms in which Art has learned to embody itself. In Painting they have done some things that are worthy of praise ; in Sculpture they have done very little ; and in Architectural Art they certainly have not achieved success. Notwi'hstanding that they have a climate inviting to architectural display in every form ; though they have the best of materials in infinite abundance ; though they had wealth and learning, and were stimulated by the example of what had been done in their own country, and was doing by other nations—in spite of all this, they have fallen far short of what was effected either in Italy or France, and now seem to be utterly incapable of appreciating the excellencies of Architectural Art, or of caring to enjoy them."

FRANCE.

After the reigns of Charles VIII. and Louis XII. the French people became somewhat familiar with Italian Art, and at length, during the reign of Francis I., from 1515 to 1546, everything Italian was the fashion in France. Francis invited such artists as Leonardo da Vinci, Benvenuto Cellini, Primaticcio, and Andrea del Sarto to come to France and aid him in his works at Fontainebleau and elsewhere.

It was not long before the Gothic architecture which had been so much used and improved in France was thought to be inferior in beauty to the Italian architecture as it existed in the sixteenth century, and very soon the latter style was adopted and considered as the only one worthy of admiration. But the French architects had been so trained to the Gothic order that it was not easy for them to change their habits of design, and the result was that

new edifices were largely of the Gothic form, but were
finished and ornamented like the Italian buildings ; by this
means the effect of the whole, when completed, was such
as is seen in this picture of the church of St. Michael at
Dijon (Fig. 109).   In these days no one approves of this
union of Gothic design and Italian decoration, but when
it was the fashion it was thought to be very beautiful
by French architects.

Francis I., who was so anxious to introduce Italian art
into France, erected edifices of a very different sort from
those which he attempted to imitate.   In Italy, the principal buildings of the Renaissance were churches or con-
vents, or such as were in some way for religious uses.
Francis I. built palaces like that of Fontainebleau, and splendid châteaux like those of Chambord, or Chenonceaux, and
the Italian style of architecture could not be readily adapted
to the lighter uses of the French kings.   The splendid
massive Pitti Palace, built after the design of the great
Brunelleschi, would scarcely have harmonized with the river
banks and the lovely undulating meadows around a country
villa or château.   So it gradually happened that French
Architecture was more graceful, light, and elegant than the
architecture of the churches, monasteries, and other religious edifices of Italy, and at the same time the Italian
feeling and influence can easily be traced in the French
buildings of the time of which we speak.

In Italy the Pope and the Church governed in Art, and
considered it only as a religious means of glorifying the
Church and impressing its doctrines upon the whole people.
In France the sovereigns held the leading place, and in the
midst of their ambitions and their gayeties they found little time to consider the matter of church architecture.
Though the church of St. Eustache was erected at Paris,
and other churches were restored, it was not until 1629,
when Cardinal Richelieu ordered the building of the church

FIG. 109.—FAÇADE OF THE CHURCH OF ST. MICHAEL. *Dijon.*

finest domical edifices in Europe, and a most satisfactory example of the architecture of its class (Fig. 110).

Directly underneath this dome is the crypt in which is the sarcophagus which contains the remains of Napoleon Bonaparte. On the door which leads to the crypt are inscribed the following words, taken from the will of the exile at St. Helena: "I desire that my ashes may rest on the banks of the Seine, in the midst of the French people whom I have loved so well."

This tomb is said to have cost nearly two millions of dollars, and though it is beautiful, and in good taste in its details, yet one can but regret that all this expense should not have erected a splendid mausoleum, such as would have dignified the monumental art of France.

The church of St. Genevieve, or the Pantheon, as it is usually called, is a very important architectural work. It was twenty-six years in building, and was not completed until after the death of its architect, Soufflot, which occurred in 1781 (Fig. 111).

It is said that this church was begun as the fulfilment of a vow made by King Louis XV. when he was ill, but as the French Revolution was in progress when it was completed, it was dedicated to the "*Grands Hommes*," or the great men of France, and not to God or the sweet St. Genevieve, who was one of the patron saints of Paris.

The dome of the Pantheon is elegant and chaste, but not great in design or effect, and the whole appearance of the church is weakened by the extreme width of the spaces between the front columns; this makes the entablature appear weak, and is altogether a serious defect. Another striking fault is the way in which a second column is placed outside at each end of the portico; one cannot imagine a reason for this, and it is confusing and unmeaning in the extreme. The interior of the Pantheon is superior to the exterior, and many authorities name it as the most satis-

Fig. 112.—The Madeleine. *Paris.*

factory of all modern, classical church interiors ; when it
was built it was believed to be as perfect an imitation of
antique classical architecture as could be made, and all the
world may be grateful that it escaped the fate prepared for
it by the Communists.   This was averted by the discovery
and cutting of the fuse which they had prepared for its
destruction on May 24th, 1871 ; the fuse led to the crypts
beneath the church, where these reckless men had placed
large quantities of powder.

In the beginning of the present century French architects believed it best to reproduce exactly ancient temples
which had been destroyed.   According to this view the
church of the Madeleine was begun in 1804, after the
designs of Vignon.   Outwardly it is a temple of the Corinthian order, and is very beautiful, though its position
greatly lessens its effect.   If it were on a height, or standing in a large square by itself, it would be far more imposing (Fig. 112).

The church of the Trinity and that of the Augustines,
at Paris, are important church edifices of the present day,
but though much thought and time have been lavished on
them, they are not as attractive as we could wish the works
of our own time to be ; and they seem almost unworthy of
attention when we remember that in the same city there are
so many examples of architecture that have far more artistic
beauty, as well as the additional charms of age and the
interest of historical associations.

We have already spoken of the sort of building in which
Francis I. delighted.   Of all his undertakings the rebuilding of the Louvre was the most successful.   Its whole design was fine and the ornaments beautiful ; many of these
decorations were made after the drawings of Jean Goujon,
who was an eminent master in such sculptures.   The court
of the Louvre has never been excelled in any country of
Europe ; it is a wonderful work for the time in which it

FIG. 113.—PAVILION DE L'HORLOGE AND PART OF THE COURT OF THE
LOUVRE.

was built, and satisfies the taste of the most critical ob
servers (Fig. 113).

We cannot give space to descriptions of the châteaux
built by Francis I., but this picture of that of Chambord
affords a good example of what these buildings were
(Fig. 114).

From the time of the reign of Charles IX. (1560) to the
close of the reign of Louis XIII., the style of architecture
which was used in France was called the " style of Henry
IV. ;" this last-named king ruled before Louis XIII., and
during his time architecture sank to a very low plane—there
was nothing in it to admire or imitate.   Under Louis XIII.

it began to improve, and in the days of Louis XIV., who
is called the "*Grand Monarque*," all the arts made great
progress and received much patronage from the king, and
all the people of the court, for whom the king was a model.
Louis XIV. began a revival of Roman classical architecture, and there is no doubt that he believed that he
equalled, or perhaps excelled, Julius Cæsar and all other
Roman emperors as a patron of the Fine Arts.

But we know that this great monarch was deceived by
his self-love and by the flatteries of those who surrounded
him and wished to obtain favors from him. His architectural works had so many faults that it is very tiresome to
read what is written about them, and in any case it is
pleasanter to speak of virtues than of faults. The works
of Louis XIV. were certainly herculean, and when we
think of the building of the palace of Versailles, the completion of the Louvre, and the numberless hôtels, châteaux,
and palaces which belong to his reign, we feel sure that if
only the vastness of the architectural works of his time is
considered, he well merits the title of the Great Monarch.
But these important edifices require more-time and space if
spoken of in detail than we can give, and I pass to some
consideration of the works of our own time.

The architecture of the reign of Napoleon III. requires
the space of a volume, at least, were it to be clearly described, for during that reign there was scarcely a city of
France that did not add some important building to its
public edifices. First, the city of Paris was remodelled and
rebuilt to a marvellous extent, and as in other matters Paris
is the leader, so its example was followed in architecture.
The new Bourse in Lyons, the Custom House at Rouen,
and the Exchange at Marseilles are good specimens of what
was done in this way outside the great metropolis.

During the reign of Louis Philippe, and a little later,
French domestic architecture was vastly improved, and

FIG. 114.—CHÂTEAU OF CHAMBORD

since then much more attention has been given by Frenchmen to the houses in which they live. The appearance of the new Boulevards and streets of Paris is picturesque, while the houses are rich and elegant. Many portions of this city are more beautiful than any other city of Europe ; and yet it is true that the architecture of forty years or so ago was more satisfactory than that of the present time.

FIG. 115.—PORTE ST. DENIS. *Paris.*

The French are an enthusiastic people, and have been very fond of erecting monuments in public places which would remind them continually of the glories of their nation, the conquests of their armies, and the achievements of their great men. Triumphal Arches and Columns of Victory are almost numberless in France ; many of them are impressive, and some are really very fine in their architecture. Since the Porte St. Denis was (Fig. 115) erected, in

FIG. 116.—ARC DE L'ÉTOILE. *Paris.*

1672, almost every possible design has been used for these monuments, in one portion of France or another, until, finally, the Arc de l'Étoile (Fig. 116) was built at the upper end of the Champs Elysées, at Paris. This is the noblest of all modern triumphal arches, as well as one of the most splendid ornaments in a city which is richly decorated with architectural works of various styles and periods—from that of the fine Renaissance example seen in the west front of the Louvre, built in 1541, down to the Arc de l'Étoile, the Fontaine St. Michel, and the Palais du Trocadéro of our own time.

The French architecture of the present century is in truth a classic revival ; its style has been called the *néo-*

Grec, or revived Greek, and the principal buildings of the reign of Napoleon III. all show that a study of Greek art had influenced those who designed these edifices.

ENGLAND.

We may say that England has never had an architecture of its own, since it has always imitated and reproduced the orders which have originated in other countries. The Gothic order is more than any other the order of England, and, in truth, of Great Britain. All English cathedrals, save one, and a very large proportion of the churches, in city and country, are built in this style of architecture.

It is also true that during the Middle Ages, when the Roman Catholics were in power in England and made use of Gothic architecture, they built so many churches, that, during several later centuries, it might be truly said that England had no church architecture, because so few new churches were required or built.

It is so difficult to trace the origin and progress of the Classical or Renaissance feeling in English architecture that I shall leave it altogether, and passing the transition style and period, speak directly of the first great architect of the Renaissance in England, Inigo Jones, who was born in 1572 and died in 1653. He studied in Italy and brought back to his native country a fondness for the Italian architecture of that day. He became the favorite court architect, and there are many important edifices in England which were built from his designs. His most notable work was the palace of Whitehall, though his design was never fully carried out in it ; had it been, this palace would have excelled all others in Europe, either of earlier or later date. Among the churches designed by Inigo Jones that of St. Paul's, Covent Garden, is interesting because it is probably the first important Protestant church erected in England which still exists. It is small and simple, being almost an exact reproduction of the early Greek temples called *distyle in antis*, such as I described when speaking of Greek architecture (Fig. 117).

FIG. 117.—EAST ELEVATION OF ST. PAUL'S. *Covent Garden.*

Inigo Jones made many designs for villas and private residences, and perhaps he is more famous for these works than for any others. Among them are Chiswick and Wilton House, and many others of less importance.

After Jones came Sir Christopher Wren, who was the architect of some of the finest buildings in London. He was born in 1632 and died in 1723. The great fire, in 1666, when he was thirty-four years old, gave him a splendid opportunity to show his talents. Only three days after this fire he presented to the king a plan for rebuilding the city, which would have made it one of the most convenient as well as one of the most beautiful cities of the world.

Sir Christopher Wren is most frequently mentioned as the architect of St. Paul's Cathedral. This was commenced nine years after the great fire, and was thirty-five years in building. St. Paul's is the largest and finest Protestant cathedral in the world, and among all the churches of Europe that have been erected since the revival of Classical architecture, St. Peter's, at Rome, alone excels it (Fig. 118).

Although so many years were consumed in the building of St. Paul's, Sir Christopher Wren lived to superintend it

FIG. 118.—ST. PAUL'S, LONDON. *From the West.*

all, and had the gratification of placing the topmost stone
in the lantern of this splendid monument to his genius.

The western towers of Westminster Abbey are said to
have been built after a design by Wren, but of this there is
a doubt.   Among his other works in church architecture
are the steeple of Bow Church, London ; the church of St.
Stephen's, Walbrook ; St. Bride's, Fleet Street, and St.
James's, Piccadilly.

The royal palaces of Winchester and Hampton were designed by Wren, and many other well-known edifices, among which is Greenwich Hospital. He made some signal failures, but it is great praise to say, what is undoubtedly true, that, though he was a pioneer in the Renaissance architecture of England, and died a century and a half ago, no one of his countrymen has surpassed him, and we may well question whether any other English architect has equalled him.

FIG. 119.—ST. GEORGE'S HALL. *Liverpool.*

Churches, palaces, university buildings, and fine examples of municipal and domestic architecture are so numerous in England and other portions of Great Britain that we cannot speak of them in detail.   The culmination of the taste for the imitation of Classical architecture was reached about the beginning of the present century, and among the most notable edifices in that manner are the British Museum, Fitzwilliam College, Cambridge, and St. George's Hall, Liverpool (Fig. 119).

A revival of Gothic Architecture has taken place in England in our own time. The three most prominent secular buildings in this style are Windsor Castle, the Houses of Parliament, and the New Museum, at Oxford. Of course, in the case of Windsor Castle, the work was a remodelling, but the reparations were so extensive as to almost equal a rebuilding. Sir Jeffry Wyatville had the superintendence of it, and succeeded in making it appear

FIG. 120.—WINDSOR CASTLE.

like an ancient building refitted in the nineteenth century— that is to say, it combines modern luxury and convenience in its interior with the exterior appearance of the castellated fortresses of a more barbarous age (Fig. 120).

In the Houses of Parliament there was an attempt to carry out, even to the minutest detail, the Gothic style as it existed in the Tudor age, when there was an excess of ornament, most elaborate doorways, and the fan-tracery vaultings were decorated with pendent ornaments which look like clusters of stalactites. Sir Charles Barry was its architect. The present school of artists in England are never weary of abusing it ; they call it a horror and declare its style to be obsolete. In fact, it is not the success at which Barry aimed ; but it excels the other efforts to revive the Gothic in this day, not only in England, but in all

FIG. 121.—THE HOUSES OF PARLIAMENT. *London.*

Europe, and has many points to be admired in its plan and its detail, while the beauty of its sky-line must be admitted by all (Fig. 121).

In the New Museum of Oxford, the Gothic is that of Lombardy, rather than the Early English. It is an example of the result of the teaching of Mr. Ruskin. It does not realize the expectations of those who advocated this manner of building, and has proved a great disappointment to the advanced theorists of a quarter of a century ago.

English architecture of the present day may be concisely described by saying that it is Gothic for churches, parson-age-houses, school-houses, and all edifices in which the clergy are interested or of which they have the oversight. On the other hand, palaces, town-halls, municipal build-ings, club-houses, and such structures as come within the care of the laity, are almost without exception in the Classic style.

Neither of these orders seems to be exactly suited to the climate of England or to the wants of its people ; therefore, neither would satisfy the demands of the ancients, who taught that the architecture of a nation should be precisely adapted to its climate and to the purposes for which the edifices are intended. In fact, the ancients carried their ideas of fitness so far that one could tell at a glance the object for which a structure had been designed ; we know that it is not possible to comply with this law in this day, although it is doubtless in accord with the true ideal of what perfect architecture should be. At the present day there is little doubt that the edifices of the Church and clergy are far more praiseworthy and true architecturally than are those for secular and domestic uses.

GERMANY.

I shall not speak of the period of the Renaissance in Germany, but shall go forward to the time of the Revival of Classic Architecture, which dated about 1825. During the eighteenth century the discoveries which were made in Greece were of great interest to all the world, and the draw-ings which were made of the temples and monuments, as well as of the lesser objects of art which existed there, were sent all over Europe, and had such an effect upon the different nations, that with one accord they began to adopt the Greek style of architecture, whenever any important work was to be done. This effect was very marked in Germany, and the German architects tried to copy every detail of Greek architecture with great exactness.

When we begin to speak of modern German architecture at this point, we do not omit anything important, for the struggles of the Reformation, and the results of the Thirty Years' War were such, that no great architectural advances were attempted for a long time. Again, the division of Germany into many small principalities, and the establishment of many little courts so divided the wealth of the German people into small portions, that no one was rich enough to undertake large buildings. There was no one great central city as in France and England, and no one sovereign was rich enough to adorn his capital with splendid edifices or to be a magnificent patron of art and artists after the fashion of the *"Grand Monarque"* in France.

Before taking up the Revival, however, I wish, for two reasons, to give a picture of the Brandenburg Gate, at Berlin. This gate was erected between 1784 and 1792. It is important because such monuments are more rare in Germany than in other European countries, especially of the time in which this was built, and because it is one of the best imitations of Greek art that exists in any nation (Fig. 122).

It is interesting to remember that when Napoleon entered Berlin as a conqueror, after the Battle of Jena, he sent the Car of Victory, which surmounts this gate, to Paris, as a trophy of his prowess. After his abdication it was returned to its original position.

The effect of the German revival of Greek art is more plainly seen in Munich than in any other city. It is the capital of Bavaria, and one of its kings, Louis I., while he was young and had not yet become king, resided at Rome ; he was a passionate lover of art, and he resolved that when he came to the throne he would make his capital famous for beautiful things. Above all, he desired to imi-

FIG. 122.—THE BRANDENBURG GATE. *Berlin.*

tate all that he had most admired in the countries he had
visited, and also the art of the ancients as he knew it from
models and pictures. For this reason it happens that
Munich is a collection of copies of buildings which have
existed in other countries and in past ages, and as these
buildings, which were first made in marble and stone, are
mostly copied in plaster in Munich, much of their beauty is
lost ; and since these copied buildings are not used for the
same purposes for which the ancient ones were intended,
the whole effect of them is very far from pleasing or satisfactory. In fact, the result is just such as must always
follow the imitation of a beautiful object, when no proper
regard is paid to the use to be made of it. If, for example,
a fine copy of a light and airy Swiss châlet should be made

in the United States of America, and placed on some business street in one of our cities, and used for a bank building, we could not deny that it was an exact copy of a building which is good in its way ; but it would be so unsuited
to its position and its uses, that the man who built it there
would be counted as insane or foolish.   And this is the
effect of the modern architecture of Munich ; it seems as

FIG. 123.—THE BASILICA AT MUNICH.

if King Louis must have been a madman to expend so
much time and money in this absurd kind of imitative
architecture, and yet it is very interesting to visit this city
and see these edifices.

Of the Munich churches erected under Louis I. that of
St. Ludwig is in the Byzantine order ; the Aue Kirche is in
the pointed German Gothic, and the Basilica is like a
Roman basilica of the fifth century.   It resembles that of

St. Paul's-without-the-Walls ; it was begun in 1835 and completed in 1850. In a vault beneath this basilica Louis and his Queen, Theresa, are buried. The picture given here shows its extreme simplicity ; its whole effect is solemn and satisfactory ; still one must regret that since it is so fine up to a certain point, it should not have been made still finer (Fig. 123).

The Ruhmeshalle, or Hall of Fame, at Munich, is an interesting and somewhat unique edifice. It is a portico

FIG. 124.—THE RUHMESHALLE. *Near Munich.*

of marble with forty-eight Doric columns, each twenty-six feet high. Against the walls are brackets holding busts of celebrated Germans who have lived since 1400. In front of the portico stands the colossal bronze statue of Bavaria. She is represented as a protectress with a lion by her side ; in the right hand she holds a sword, and a chaplet in the left ; it is sixty-one and a half feet high, and the pedestal raises it twenty-eight and a half feet more ; inside, a staircase leads up into the head, where there are seats for eight persons. The view from the top of this statue is fine, and so extensive that in a favorable atmosphere the heights of the Alps can be discerned. The hill upon which the Ruhmeshalle is built is to the south of Munich, and is called the Theresienhöhe. The grand statue is intended to be the principal object of interest here, and the portico is made so low as to throw the figure out and show it off to advantage ; altogether it is one of the most successful architectural works in Munich (Fig. 124).

The Glyptothek, or Sculpture Gallery, the Pinakothek, or Picture Gallery, the Royal Palace, the Public Library,

FIG. 125.—THE MUSEUM. *Berlin.*

the War Office, the University, Blind School, other palaces and secular buildings, all belong to the time of the Revival in Germany. The Ludwig Strasse, which King Louis fondly hoped to make one of the most beautiful avenues in the world, is—with its Roman arch at one end, and a weak copy of the Loggia dei Lanzi at the other—a tiresome, meaningless, architectural failure.

The Museum of Berlin is a striking result of the same Revival of Classic architecture, and is far more splendid than anything in Munich (Fig. 125).

In Dresden the most important works in this style are the New Theatre and Picture Gallery. The last is almost an exact reproduction of the Pinakothek of Munich. All over Germany the effects of this Revival are more or less prominent, but I shall speak of but one other edifice, the Walhalla (Fig. 126).

This is also a Temple of Fame, and is situated about six miles from Ratisbon. It overlooks the River Danube from a height of more than three hundred feet. It was begun in 1830, and was twelve years in building, costing eight millions of florins. It is of white marble, and on the exterior

FIG. 126.—THE WALHALLA.

is an exact reproduction of the Parthenon at Athens. The interior is divided into two parts by an entablature, which supports fourteen caryatides, made from colored marbles. These figures in turn support a second entablature, on which is a frieze in eight compartments, on which is sculptured scenes representing the history of Germany from its early days to the time of the introduction of Christianity. Along the lower wall there are one hundred busts of illustrious Germans who had lived from the earliest days of Germany down to those of the poet Goethe.

The grounds about the Walhalla are laid out in walks, and from them there are fine, extensive views. Taken by itself there is much to admire in the Walhalla. The sculptures arouse an enthusiasm about Germany, her history, and the men who have helped to make it, in spite of the strange unfitness with which the artists have mingled Grecian myths and German sagas. But aside from this sort of interest the whole thing seems incongruous and strangely unsuited to its position ; one writer goes so far as to say of it that " Minerva, descending in Cheapside to separate two quarrelling cabmen, could hardly be more out of place." And yet it is true that the Walhalla is the only worthy rival to St. George's Hall, Liverpool, as an example of the possible adaptability of Greek or Roman Architecture to the needs and uses of our own days.

THEATRES AND MUSIC HALLS.

In speaking of theatres I will first give a list of the most important ones in Europe, as they are given by Fergusson in his " History of Modern Architecture."

	Depth from Curtain to back of Boxes.	Depth of Stage.
	feet.	feet.
La Scala, Milan....................	105	77
San Carlo, Naples..................	100	74
Carlo Felice, Genoa................	95	80
New Opera House, Paris............	95	98
Opera House, London (old).........	95	45
Turin Opera House	90	110
Covent Garden, London	89	89
St. Petersburg, Opera.............	87	100
Académie de Musique, Paris	85	82
Parma, Opera.....................	82	76
Fenice, Venice	82	48
Munich Theatre...................	80	87
Madrid Theatre...................	79	55

The Opera House of La Scala, at Milan, is generally said to be the finest of all for seeing and hearing what goes on upon the stage : it was begun in 1776 and finished two years later.   San Carlo, Naples, holds the second place, and was first erected in 1737, but was almost destroyed by fire in 1816, and was afterward thoroughly rebuilt.

The new Opera House of Paris is interesting to us because it has been built so recently and so much written and said of it that we are familiar with it.   Any description

FIG. 127.—THE NEW OPERA HOUSE. *Paris.*

that would do it justice would occupy more space than we can afford for it, but this cut (Fig. 127) gives an excellent idea of its size and exterior appearance.   It is distinguished by great richness of material and profusion of ornament, its interior decorations being especially splendid.   It has been criticised as lacking repose and dignity, but its elegance and magnificence compel admiration.

Music halls are only another sort of theatre, and have come into great favor in recent days, especially in England.

The Albert Hall, South Kensington, is the finest music hall
that has been erected. It seats eight thousand people,
besides accommodating an orchestra of two hundred and a
chorus of one thousand singers ; it is one hundred and
thirty-six feet from the floor to the highest part of the ceiling. This hall has some defects, but is so far successful as
to prove that a theatre or music hall could be so constructed
as to seat ten thousand persons and permit them to hear
the music as distinctly as it is heard in many halls where
only two or three thousand can be comfortable.

UNITED STATES OF AMERICA.

When we remember that we have been able to give
some account of architecture as it existed thousands of
years before Christ, and to speak of the temples and
tombs of the grand old nations who laid the foundation of
the arts and civilization of the world—and then, when we
remember the little time that has passed since the first roof
was raised in our own land, we may well be proud of our
country as it is—and at the same time we know that its
architecture may in truth be said to be a thing of the
future.

It is but a few years, not more than seventy, since any
building existed here that could be termed architectural in
any degree. To be sure, there were many comfortable,
generous-sized homes scattered up and down the land, but
they made no claim to architectural design, and were not
such edifices as one considers when speaking or writing of
architecture.

The first buildings to which much attention was given
in the United States were the Capitols, both State and
National, and until recently they were in what may be
called a Classic style, because they had porticoes with columns and certain other features of ancient orders ; but when

FIG. 128.—THE UNITED STATES CAPITOL. *Washington.*

the cella, as is the case in America, is divided into two or more stories, with rows of prosaic windows all around, and chimneys, and perhaps attics also added, the term Classic Architecture immediately becomes questionable, and it is difficult to find a name exactly suited to the needs of the case ; for it is still true that from a distance, and in answer to a general glance, they are nearer to the Classic orders than to anything else.

The National Capitol at Washington, which is the principal edifice in the United States, was begun in 1793, when General Washington laid the foundation-stone ; the main portion was completed in 1830 ; two wings and the dome have since been added, and its pres-

ent size is greater than that of any other legislative building in the world, except the British Houses of Parliament
(Fig. 128).

The dome, and the splendid porticoes, with the magnificent flights of steps leading up to them, are the fine features of the Capitol. The dome compares well with those
that are famous in the world, and taken all in all the Wash-

FIG. 129.—STATE CAPITOL. *Columbus, Ohio.*

ington Capitol is more stately than the Houses of Parliament, and is open to as little criticism as buildings of its
class in other lands.

Several of the State Capitols illustrate the manner of
building which I described above. This cut of the Capitol
of Ohio is an excellent example of it (Fig. 129).

In domestic architecture, while there has been no style
so original and absolutely defined as to be definitely called

American, we may roughly classify three periods—the Colonial, the Middle, and the Modern. These terms have no close application, and you must understand that I use them rather for convenience than because they accurately, or even approximately, indicate particular styles. The mansions of the Colonial period are, perhaps, most easily recognized, and in some respects were the frankest and most independent class of houses ever built in this country. The early settlers took whatever suited them from all styles, and instead of imitating the English, the Dutch, or the French manner of building, mingled parts of all, with especial reference to the needs of their climate and surroundings.

This fine old house (Fig. 130) shows the plain, homely, yet quaint style of many of the mansions of the Colonial period. It was built near the beginning of the last century, and occupied by Sir William Pepperell until his death. Its interior, with heavy wainscoting of solid mahogany, was more imposing by far than the exterior. The Van Rensselaer homestead at Albany is an excellent example of a more stately house, possessing much dignity and impressiveness.

The Middle period was a time when domestic architecture, still without any originality and losing much of the independence of the Colonial, copied more closely from foreign models. Some fine old mansions belong to this period, which covered the last years of the last century and the first half of this. The celebrated Cragie House at Cambridge, occupied by the poet Longfellow ; " Elmwood," the home of James Russell Lowell ; " Bedford House," in Westchester County, New York, the home of the Hon. John Jay, are to be referred to this period ; and so is the imposing " Old Morrisania," at Morrisania, New York, the old Morris mansion (Fig. 131).

It is modelled after a French château, and was erected by General Morris after his return from France in 1800. It

FIG. 130.—SIR WILLIAM PEPPERELL'S HOUSE, Kittery Point, Maine.

is one of the most striking among the mansions of its time, and both its interior and exterior are highly interesting.

These views serve to illustrate the want of anything like a regular style, of which I spoke above ; but they show how many different forces were at work to influence building in the Modern period. This division is meant to extend to and include the present time, and so great is the diversity of styles now employed that in a work like this it would be idle to attempt anything like an enumeration of them, and still less to try and determine their origin and importance. I can only give you one example of the handsome and costly homes which are being built to-day, and leave you to observe others as you now see them everywhere about the country (Fig. 132). A modern writer on American architecture claims that in private dwellings an American order is gradually being developed by the changes made to adapt foreign forms to our climate, and especially to the brilliancy of the sunlight here. All this is so difficult to define, however, that it would be impossible to show it clearly in the limits of a book like this, even if it exists.

What is called the " Queen Anne" style, modelled upon the English fashion of the time of that monarch, is very widely used in country houses at the present time, sometimes in conjunction with the Colonial, which also exists as an independent style. The tendency of domestic architecture is to make everything quaint and picturesque, though this is not so far carried to extremes as was the case a few years since.

In public buildings many splendid edifices have been erected of late years. The imitation of classic forms which was formerly the fashion, and which is so strikingly exhibited by Girard College, Philadelphia, is now almost entirely laid aside. A lighter, less constrained style, which may be called eclectic—which means selecting—because it takes freely from any and all styles whatever suits its pur-

FIG. 131.—"OLD MORRISANIA." *Morrisania, New York.*

pose, is arising ; and as this selecting is being every year
more and more intelligently done, and as original ideas are
constantly being incorporated with those chosen, the pros-
pects for architecture are more promising than ever before
in this country.   The Casino, at Newport, is a fine example
of a modern building ; and the still more recent Casino in
New York shows a fine example of the adapting of ideas
from Saracenic architecture to American uses.   The Capitol
at Albany has many fine features, but it is the work of
several designers who did not harmonize.   Memorial Hall,
at Cambridge, is one of the more striking of modern American buildings, but its sky-line--that is, its outline as seen
against the sky—lacks simplicity and repose.

The churches in this country exhibit the widest variety
of style.   Trinity Church in New York was the first Gothic
church erected in America, and Trinity Church in Boston, one
of the latest churches of importance, is also Gothic, though of
the variety called Norman Gothic, and considerably varied.
The Roman Catholic Cathedral of New York, and many
others of less magnitude, might be cited as a proof that
American architecture is advancing, and that we may speak
hopefully of its future.

Railroad depots and school-houses of certain types are
among the most distinctive and characteristic American
edifices.   The first, especially, are being constructed more
nearly in accordance with the ancient principle of suiting
the structure to its uses than are any other buildings that
are worthy to be considered architecturally.   Art museums
and public libraries, too, now form an important feature
in both town and country, and, in short, the beginning of
American architecture, for that is all that can be claimed
for what as yet exists, is such as would be the natural out-
come of a nation such as ours—varied, restless, bold, ugly,
original, and progressive.   All these terms can be applied
to American art, but in and through it all there is a prom-

FIG. 132—RESIDENCE AT IRVINGTON, NEW YORK.

ise of something more. As greater age will bring repose
and dignity of bearing to our people, so our Fine Arts will
take on the best of our characteristics ; as we outgrow our
national crudities the change will be shown in our architecture, and we may well anticipate that in the future we shall
command the consideration and assume the same importance in these regards that our excellence in the Useful Arts
has already won for us in all the world.

GLOSSARY OF TECHNICAL TERMS.

Abacus.—The uppermost portion of the capital of a column, upon which rested the weight above.

Aisle.—The lateral divisions of a church ; more properly, the side subdivisions.

Amphitheatre.—A round or oval theatre.

Apse.—The semi-circular or polygonal termination to the choir or aisles of a church.

Arcade.—A series of arches supported on piers or columns.

Arch.—A construction of wedge-shaped blocks of stone or of bricks, of curved outline, spanning an open space.

Architrave.—(1) The lowest division of the entablature, in Classic architecture resting on the abacus. (2) The moulding used to ornament the margin of an opening.

Base.—The foot of a column or wall.

Basilica.—Originally a Roman hall of justice ; afterward an early Christian church.

Buttress.—A projection built from a wall for strength.

Byzantine.—The Christian architecture of the Eastern church, sometimes called the round arched ; named from Byzantium (Constantinople).

Capital.—The head of a column or pilaster.

Caryatid.—A statue of a woman used as a column.

Cathedral.—A church containing the seat of a bishop.

Cella.—That part of the temple within the walls.

Chamfer.—A slope or bevel formed by cutting off the edge of an angle.

Column.—A pillar or post, round or polygonal ; the term includes the base, shaft, and capital.

Composite Order.—See *Order.*

Corinthian Order.—See *Order.*

Cornice.—The horizontal projection crowning a building or some portion of a building. Each classic order had its peculiar cornice.

Crypt. —A vault beneath a building.

Dome.—A cupola or spherical convex roof.

Doric Order.—See *Order.*

Entablature.—In classic styles all the structure above the columns except the gable. The entablature had three members, the architrave or epistyle, the frieze, and the cornice.

Entasis.—The swelling of a column near the middle to counteract the appearance of concavity caused by an optical delusion.

Epistyle.—See *Architrave.*

Façade.—The exterior face of a building.

Frieze.—The middle member of an entablature.

Gable.—The triangular-shaped wall supporting the end of a roof.

Gargoyle.—A projecting water-spout carved in stone or metal.

Hexastyle.—A portico having six columns in front.

Intercolumniation.—The clear space between two columns.

Ionic Order.—See *Order*.

Metope.—The space between the triglyphs in the frieze of the Doric Order.

Minaret.—A slender tower with balconies from which Mohammedan hours of prayer are called.

Mosaic.—Ornamental work made by cementing together small pieces of glass, stone, or metal in given designs.

Nave.—The central aisle of a church ; the western part of the church occupied by the congregation.

Obelisk.—A quadrangular monolith terminating in a pyramid.

Order.—An entire column with its appropriate entablature. There are usually said to be five orders : Tuscan, Doric, Ionic, Corinthian, and Composite ; the first and last are, however, only varieties of the Doric and Corinthian developed by the Romans. The peculiarities of the orders have been described in the body of the book. When more than one order was used in a building, the heavier and plainer, the Doric and Tuscan, are placed beneath the others.

Pediment.—In classic architecture what the gable (which see) was in later styles.

Peristyle.—A court surrounded by a row of columns ; also the colonnade itself surrounding such a space.

Pier.—A solid wall built to support a weight.

Pilaster.—A square column, generally attached to the wall.

Pillar.—See *Column*.

Plinth.—A square member forming the lower division of the base of a column.

Polychrome.—Many-colored ; applied to the staining of walls or architectural ornaments.

Quatrefoil.—A four-leaved ornament or opening.

Shaft.—The middle portion of a column, between base and capital.

Story.—The portion of a building between one floor and the next.

Triglyph.—An ornament upon the Doric frieze consisting of three vertical, angular channels separated by narrow, flat spaces.

INDEX.